W0091936

Handbook of Natural Gas Transmission and Processing

Handbook of Natural Gas Transmission and Processing

Editor

Ashish Vishnoi

Handbook of Natural Gas Transmission and Processing
Edited by **Ashish Vishnoi**

Printed in 2017

ISBN: 978-1-68117-394-8

Library of Congress Control Number: 2015941583

© 2016 by

SCITUS Academics LLC,
616, Corporate Way, Suite 2, 4766,
Valley Cottage, NY 10989

www.scitusacademics.com

This book contains information obtained from highly regarded resources. Copyright for individual articles remains with the authors as indicated. All chapters are distributed under the terms of the Creative Commons Attribution License, which permits unrestricted use, distribution, and reproduction in any medium, provided the original author and source are credited.

Notice

Reasonable efforts have been made to publish reliable data and views articulated in the chapters are those of the individual contributors, and not necessarily those of the editors or publishers. Editors or publishers are not responsible for the accuracy of the information in the published chapters or consequences of their use. The publisher believes no responsibility for any damage or grievance to the persons or property arising out of the use of any materials, instructions, methods or thoughts in the book. The editors and the publisher have attempted to trace the copyright holders of all material reproduced in this publication and apologize to copyright holders if permission has not been obtained. If any copyright holder has not been acknowledged, please write to us so we may rectify.

Contents

Preface

Handbook of Natural Gas Transmission and Processing gives engineers and managers complete coverage of natural gas transmission and processing in the most rapidly growing sector to the petroleum industry. The authors provide a unique discussion of new technologies that are energy efficient and environmentally appealing at the same time. Handbook of Natural Gas Transmission and Processing provide a fresh look at new technologies and opportunities for solving current gas processing problems on plant design and operation and on greenhouse gases emissions. It also does an excellent job of highlighting the key considerations that must be taken into account for any natural gas project in development.

Editor

Theoretical Study of Hydrogen Bond Formation in Trimethylene Glycol-Water Complex

Snehanshu Pal and T. K. Kundu

Department of Metallurgical and Materials Engineering, Indian Institute of Technology Kharagpur, West Bengal, Kharagpur 721302, India

ABSTRACT

A detailed quantum chemical calculation based study of hydrogen bond formation in trimethylene glycol- (TMG-) water complex has been performed by Hatree-Fock (HF) method, second-order Møller-Plesset perturbation theory (MP2), density functional theory (DFT), and density functional theory with dispersion function (DFT-D) using 6-31++G(d,p) basis set. B3LYP DFT-D, WB97XD, M06, and M06-2X functionals are used to capture highly dispersive hydrogen bond formation. Geometrical parameters, interaction energy, deviation of potential energy curve of hydrogen-bonded O–H from that of free O–H, natural bond orbital (NBO), atom in molecule (AIM), charge transfer, and red shift are investigated. It is observed that hydrogen bond between TMG and water molecule is stronger in case of TMG acting as proton donor compared to that of water acting as proton donor, and dilute TMG solution would inhibit water cluster formation.

INTRODUCTION

A hydrogen bond is an attractive donor-acceptor interaction, in which generally, the donor atoms are electronegative compared to hydrogen, and acceptor atoms have unshared lone pair electrons [1, 2]. The hydrogen bond has crucial impacts on many aspects of chemical and biological systems, and accordingly, hydrogen bond interaction has been an important research topic for several decades. The hydrogen bond also plays a significant role in formation of clathrate hydrate in marine sediments and below permafrost regions, which is considered to be significant future energy source [3, 4]. Global warming due to aleatory decomposition of methane hydrate [5] and hazards in petroleum industry owing to formation of gas hydrate in oil pipe line are of great concern [6, 7]. Controlled inhibition of gas hydrate formation is thus, very important, and various thermodynamic and kinetic inhibitors can break hydrogen-bonded network of clathrate structure by forming itself comparatively stronger hydrogen bond with water molecules of clathrate. Knowledge of hydrogen bond interaction is essential to identify potential gas hydrate inhibitor and design effective gas hydrate inhibitor. Trimethylene glycol being a polar compound can be a potential gas hydrate inhibitor as well as antifreeze reagent [7, 8].

Ab initio calculation is one of the most appropriate ways to obtain perspicuous understanding of hydrogen bond interaction and its impact on gas hydrate inhibition. Density functional theory (DFT) and atom in molecule (AIM) study of strong dihydrogen bonds [10] and resonance-assisted hydrogen bond [11] have been performed. Several theoretical studies of hydrogen bond interaction have been carried out for different systems like water complex [12, 13], dichlorine monoxide-hydroxyl radical system [14], tetrahydrofuran-water complex [15, 16], and methanol-water complex [17]. In one word, the literature of quantum chemical analysis of hydrogen bond interaction for various complex is well enriched [18–23]. Electronic structure-based studies on hydrogen bond formation between a molecule having two hydroxyl groups (like trimethylene glycol) and water have not been reported in the literature. Explicit study of interaction between trimethylene glycol and water is necessary to reveal the effect of a molecule having two hydroxyl groups on intermolecular and intramolecular hydrogen bond formation possibilities. This electronic structure-based insights on hydrogen bond formation can help in scientific understanding on application of trimethylene glycol as a gas hydrate inhibitor.

Our objective is to report a detailed theoretical analysis to comprehend the electronic nature of the hydrogen bond formation in trimethylene glycol-water system and its property using Hartree Fock, Møller-Plesset truncated at second-order (MP2), density function theory (DFT), and density functional theory with dispersion function (DFT-D). This study will help to conceptualize the nitty-gritty of hydrogen bond formation and its effect on vibrational spectra, natural bond orbital in trimethylene glycol-water complex.

COMPUTATIONAL DETAIL

Geometry optimization, determination of interaction energy, and natural bond orbital (NBO) analysis have been carried out using Hatree Fock (HF) [24] method, second-order Møller-Plesset perturbation theory (MP2) [25], density functional theory (DFT) [26, 27], and density functional theory with dispersion function (DFT-D) [28]. The calculations for DFT and DFT-D levels of theory have been performed using different functional, namely, B3LYP [29, 30], WB97XD [31], M06 [32], and M062X [32]. As polarity [33], of molecule has great influence on intermolecular hydrogen bonding, hydrogen bond-forming orbitals require larger space occupation [34]. Thus, diffuse and polarization functions augmented split valence 6-31++G(d,p) basis set is used for better description of molecular orbitals for geometry optimization and NBO analysis. Frequency calculation as well as AIM analysis have been performed using WB97XD/6-31++G(d,p) level of theory. Since hydrogen bonding is a kind of van der Waals type interaction, additional dispersion function with density functional theory, that is, DFT-D-based calculation has also been performed.

Interaction energy (ΔE_{HBF}) for hydrogen-bonded complex is calculated as the difference between the energy of hydrogen-bonded complex and the summation of the energies of each component monomer [35] as given in (1),

$$\Delta E_{HBF} = E_{complex} - \sum E_{component},$$

(1)

where $E_{complex}$ and $E_{component}$ are optimized energy of hydrogen-bonded complex and each individual component monomer, respectively. Interaction energies have corrected for the basis set superposition error (BSSE) by virtue of counterpoise method [36]. A hydrogen-bonded complex is more stable if interaction energy is more negative compared to other hydrogen-bonded configurations.

Donor-acceptor interaction strength between filled orbital of the donor $\left(\Phi_i\right)$ and the empty orbital of acceptor $\left(\Phi_j\right)$ in case of natural bond orbital (NBO) analysis [37, 38] has been determin ed by second-order perturbation energy ($\Delta E_{ij}^{(2)}$) calculated using (2),

$$\Delta E_{ij}^{(2)} = 2\frac{\left\langle \phi_i \left| F_{ij} \right| \phi_j \right\rangle^2}{\varepsilon_i - \varepsilon_j},$$

(2)

where ε_i and ε_j are NBO energies, and F_{ij} is Fock matrix element between the i and j NBO orbitals. NBO analysis reveals the intra- and intermolecular interactions, and it is one of the appropriate methods for investigating hyperconjugative interactions [39].

Red shift in vibrational spectroscopy of conventional hydrogen-bonded structures arises from hyper-conjugation interaction [40]. Atom in molecule (AIM) study using Bader theory [41] has been performed as it is very effective for evaluating topological parameters of hydrogen bonds.

All the calculations have been carried out using Gaussian 09 software package [42]. Discovery Studio v3.1 of Accelrys software inc. is used for visualization of molecules. Vibrational frequency is calculated using 0.975 scaling factor [43].

RESULT AND DISCUSSION

Three possible conformations of trimethylene glycol denoted by TMG-1, TMG-2, and TMG-3 have been optimized using WB97XD/6-31++G(d,p) and shown in Figure 1. The TMG-2 conformation is found to be most stable as calculated relative energies of TMG-1 (3.77 kcal/mol) and TMG-3 (3.14 kcal/mol) isomers with respect to TMG-2 isomer are positive. This is because of intramolecular hydrogen bond formation in TMG-2 conformation. Detailed study of hydrogen bond interaction between water and TMG-2 conformation has been reported in this paper, and TMG-2 conformation is described as TMG. The optimized structures of TMG dimer, water dimer, TMG and one water complex considering either TMG or water as a proton donor (referred as TD and WD, resp.), and TMG + two water complex using WB97XD/6-31++G(d,p) calculation are shown in Figure 2. Here it is observed that intramolecular hydrogen bond distance (AO...O12···H5) of TMG molecule increases in presence of water in WD conformation. It is also found that intermolecular hydrogen bond distances between TMG and water are less than intramolecular hydrogen bond distance (O12···H5) for TMG(TD)-one water complex and TMG + 2 water complexes. The intermolecular hydrogen bond in TMG + 1 water complex (TD) is smaller in length and consequently stronger than that of water dimer as evident in Figure 2. The calculated structural parameters using 6-31++G(d,p) basis set and different levels of theory are summarized in Tables 1(a) and 1(b). It is identified that B3LYP DFT-D [44], parameterized functional such as M06-, M06-2x-, and WB97XD-based methods which consider attractive dispersion force, show shorter hydrogen bond distances compared to HF theory-based calculation for all the systems. It is observed from Table 1(a), hydrogen bond angle values for intramolecular

hydrogen bonds ($A_{o....H-o}$, $O12\cdots H5-O4$) of TMG molecule are less than hydrogen bond angle for intermolecular hydrogen bonds between TMG and water molecule in TMG + n water complex (n=1,2) for all calculation methods used in this paper. It is inferred based on hydrogen bond angle that the strength of intramolecular hydrogen bonds of TMG molecule are less compared to the strength of intermolecular hydrogen bonds between TMG and water molecule in TMG − water complex (n=1,2). The systems based on their dipole moment values in ascending order are water dimer < TMG < TMG + 1 water complex (TD) < TMG + 1 water complex (WD) < TMG + 2 water complex, for all the calculation procedures performed in this work, as evident in Table 1(a). Stronger intermolecular hydrogen bond formation enhances the dipole moment as hydrogen bond formation helps superposition of O⋯H moment and delocalization of π electrons in hydrogen-bonded molecular complex [45].

Table 1: (a) Calculated hydrogen bond distances ($d_{o···H'}$, Å), hydrogen bond angles ($A_{o···H'}$, degree), dipole moment (D, debye) for single TMG, and TMG – n water complex (n=1,2) using 6-31++G(d,p) basis set and various methods. (b) Calculated hydrogen bond distances ($d_{o···H'}$, Å), hydrogen bond angles ($A_{o···H'}$, degree), dipole moment (D, debye) for water dimer, and TMG dimer using 6-31++G(d,p) basis set and various methods

(a)

System	Parameters	Methods						
		MP2	WB97XD	MO6-2X	B3LYP DFTD	MO6	B3LYP	HF
TMG	$d_{o···H'}$ O12···H5	2.01	2.02	2.02	2.03	2.04	2.04	2.12
	$A_{o···H'}$ O12···H5-O4	137.17	137.36	135.62	138.00	136.10	137.45	133.52
	D	4.13	3.85	3.82	3.82	3.78	3.82	3.90
TMG + 1 water complex (TD)	$d_{o···H'}$ O14···H13	1.89	1.87	1.88	1.85	1.90	1.89	2.01
	$d_{o···H'}$ O12···H5	1.96	1.97	1.98	1.97	1.98	1.98	2.08
	$A_{o···H'}$ O14···H13-O12	179.34	179.81	172.01	178.91	166.1	177.74	179.73
	$A_{o···H'}$ O12···H5-O4	139.93	140.21	137.76	141.10	139.0	140.29	135.43
	D	5.45	5.14	5.61	4.96	5.77	5.04	5.35

TMG + 1 water complex (WD)							
$d_{O \cdots H'}$ O12···H14	2.15	2.15	2.13	2.18	2.18	2.23	2.27
$d_{O \cdots H'}$ O4···H16	2.29	2.18	2.16	2.09	2.15	2.21	2.46
$d_{O \cdots H'}$ O12···H5	2.08	2.09	2.08	2.08	2.08	2.11	2.19
$A_{O \cdots H'}$ O12···H14–O15	153.46	150.23	150.60	145.10	147.59	149.53	156.83
$A_{O \cdots H'}$ O4···H16–O15	137.16	140.56	138.92	145.31	142.07	141.43	132.04
$A_{O \cdots H'}$ O12···H5–O4	129.04	129.55	127.96	131.17	129.46	129.48	126.30
D	5.84	5.56	5.44	5.58	5.51	5.67	5.55
TMG + 2 water complex							
$d_{O \cdots H'}$ O12···H16	1.93	1.88	1.92	1.85	1.92	1.92	2.07
$d_{O \cdots H'}$ O14···H18	2.03	1.98	2.00	1.96	2.01	2.01	2.16
$d_{O \cdots H'}$ O4···H19	2.07	2.04	2.03	2.01	2.04	2.08	2.19
$d_{O \cdots H'}$ O12···H5	2.08	2.09	2.05	2.10	2.06	2.13	2.20
$A_{O \cdots H'}$ O12···H16–O14	167.66	171.38	159.54	170.48	160.10	168.10	170.42
$A_{O \cdots H'}$ O14···H18–O17	159.62	160.19	161.95	160.01	161.85	160.78	158.30
$A_{O \cdots H'}$ O4···H19–O17	153.14	152.70	151.64	154.50	153.11	154.97	154.52
$A_{O \cdots H'}$ O12···H5–O4	134.18	134.34	133.78	134.72	134.33	133.69	129.97
D	8.09	7.58	7.10	7.43	7.09	7.60	7.90

System	Parameters	Methods						
		MP2	WB97XD	MO6-2X	B3LYP DFTD	MO6	B3LYP	HF
Water dimer	$d_{o\cdots H'}$ O1···H6	1.98	1.99	1.99	1.98	1.98	1.98	2.01
	$A_{o\cdots H'}$ O1···H6–O2	175.61	175.04	172.63	174.40	175.05	174.21	176.21
	D	3.29	3.01	2.98	2.91	3.09	3.04	3.28
TMG dimer	$d_{o\cdots H'}$ O12···H5	2.0	2.01	1.89	1.97	1.89	2.04	2.12
	$d_{o\cdots H'}$ O25···H13	1.79	1.81	1.81	1.74	1.84	1.81	1.95
	$d_{o\cdots H'}$ O4···H18	—	—	1.81	—	1.84	—	—
	$d_{o\cdots H'}$ O12···H18	1.95	2.00	—	2.00	—	1.96	2.15
	$d_{o\cdots H'}$ O17···H26	—	2.34	1.89	2.31	1.90	2.49	—
	$A_{o\cdots H'}$ O12···H5–O4	136.76	136.49	144.21	140.60	144.81	139.91	133.99
	$A_{o\cdots H'}$ O25···H13–O12	158.08	156.32	157.47	161.39	155.67	157.50	158.65
	$A_{o\cdots H'}$ O4···H18–O17	—	—	157.46	—	155.76	—	—
	$A_{o\cdots H'}$ O12···H18–O17	158.78	156.74	—	143.10	—	155.14	163.54
	$A_{o\cdots H'}$ O17···H26–O25	—	114.80	144.19	116.28	144.77	108.57	—
	D	3.22	2.65	0.0011	1.31	0.0013	1.67	3.84

(b)

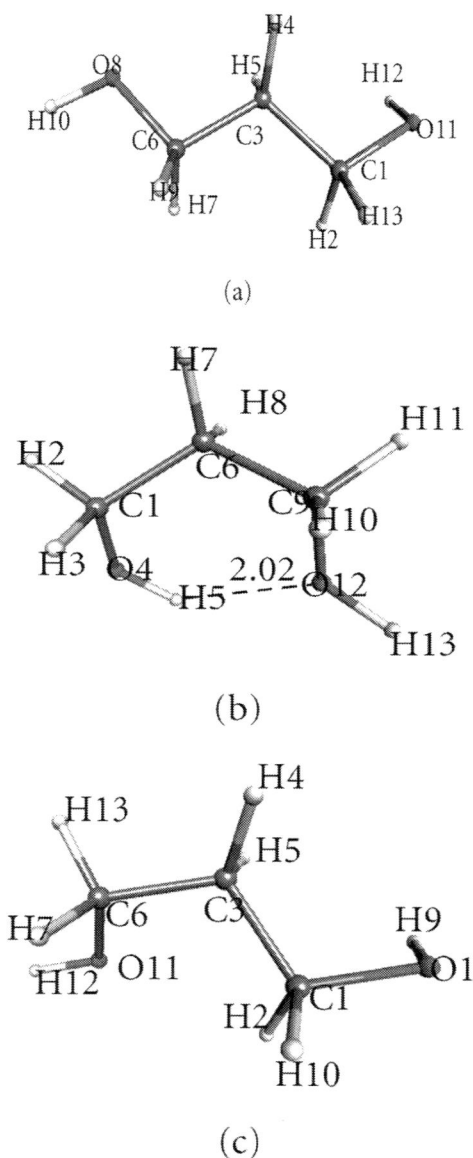

(a)

(b)

(c)

Figure 1: Optimized structures of different trimethylene glycol conformations such as (a) TMG-1, (b) TMG-2, and (c) TMG-3 using WB97XD/6-
31++G(d,p) (colour legend: red = oxygen, black = carbon, and whitish
grey = hydrogen, and black dotted line is hydrogen bond and hydrogen
bond distance in Å).

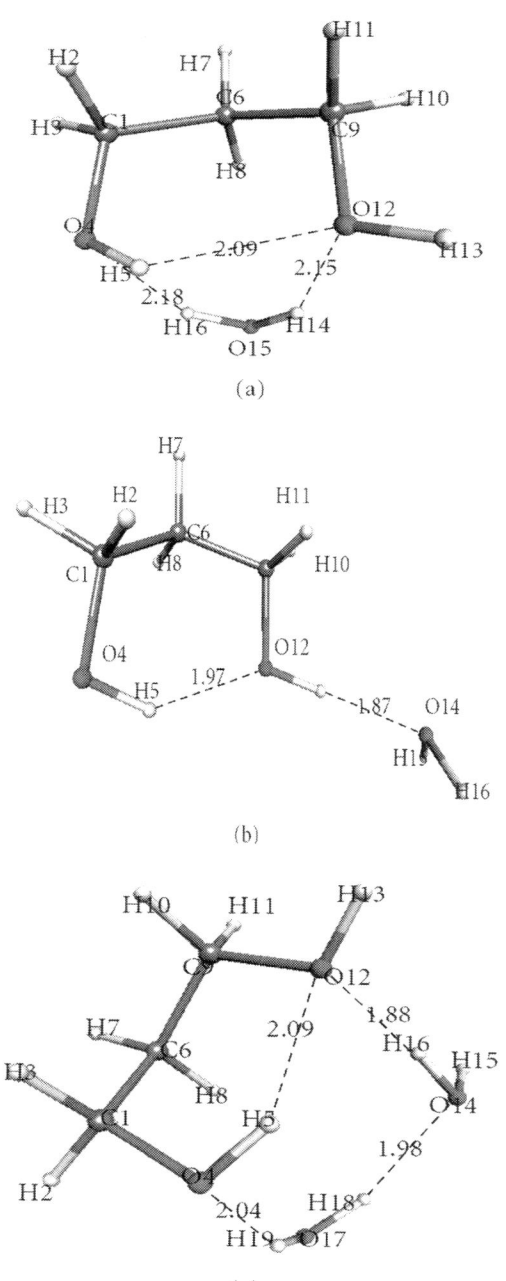

(a)

(b)

(c)

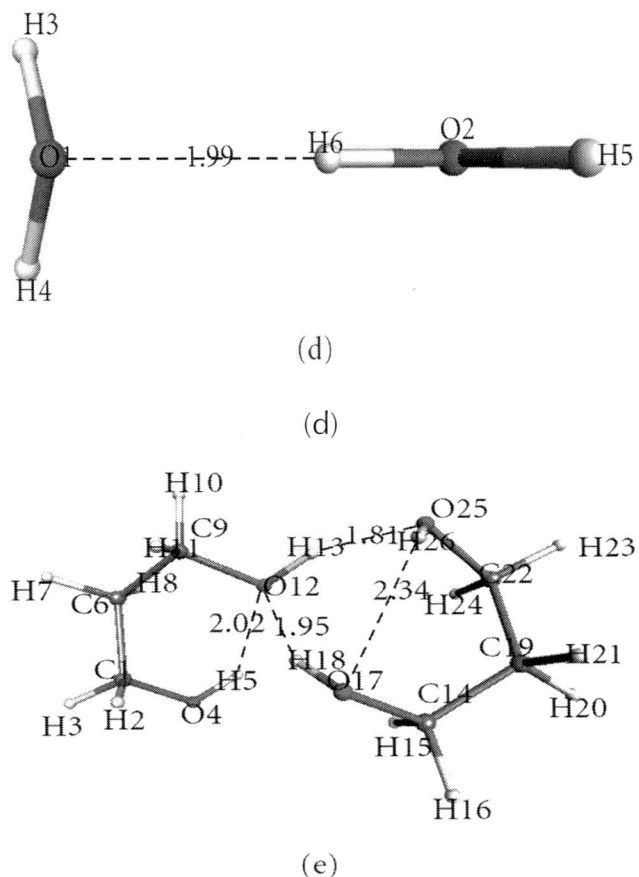

(d)

(d)

(e)

Figure 2: Optimized structures using WB97XD/6-31++G(d,p) of (a) tri-methylene glycol (TMG) + 1 water complex (WD), (b) trimethylene glycol (TMG) + 1 water complex (TD), (c) trimethylene glycol (TMG) + 2 water complex, (d) water dimer, and (e) trimethylene glycol dimer (colour legend: red = oxygen, black = carbon, and whitish grey = hydrogen, and black dotted line is hydrogen bond and hydrogen bond distance in Å).

Calculated interaction energies with $\left(\Delta E_{HBF,CP}\right)$ and without $\left(\Delta E_{HBF}\right)$ basis set superposition error (BSSE) correction (using counterpoise method) for TMG + water complexes (n=1,2), TMG dimer, and water dimer along with number of hydrogen bonds formed are summarized in Table 2. The calculated interaction energies using HF method is less negative, indicating least hydrogen bond strength as it does not consider electron correlation. The hydrogen-bonded complex having more negative interaction energy should be more stable. Therefore, the ascending order with respect to stability is TMG + 1 water (WD) complex < TMG + 1 water (TD) complex < TMG + 2 water complex using HF, MP2, and B3LYP functional based methods, which exclude the dispersion term. The stability order in ascending sense using B3LYP DFT-D, WB97XD, M06, and M06-2X functional is water dimer < TMG + 1 water (TD) complex < TMG + 1 water (WD) complex < TMG dimer < TMG + 2 water complex. TMG + 1 water complex (TD) forms one intermolecular hydrogen bond (O14···H13), and TMG + 1 water (WD) complex form two intermolecular hydrogen bonds (O12···H14, O4···H16) as shown in Figures 2(b) and 2(c). Calculation methods using functionals having dispersion terms (B3LYP DFT-D, WB97XD, M06, and M06-2X) determine more negative interaction energy for a hydrogen-bonded complex compared to that obtained by HF, MP2 and B3LYP method as evident from Table 2. TMG has strong potential to form stable cluster with water molecules, and accordingly, dilute TMG solution would be useful as an inhibitor to restrict the formation of water cluster.

Table 2: Calculated interaction energy without correction (ΔE_{HBF}, kcal/mol), BSSE-corrected energy of hydrogen bond formation using counterpoise correction ($\Delta E_{HBF,CP}$, kcal/mol), hydrogen bond numbers for TMG + n water complex (n=1,2), TMG dimer, and water dimer using 6-31++G(d,p) basis set and various methods

Systems	Calculation methods	ΔE_{HBF}	$\Delta E_{HBF,CP}$	No. of hydrogen bonds

TMG + 1 water complex (TD)	MP2	−7.65	−5.62	2
	HF	−5.64	−4.94	
	B3LYP	−6.71	−5.77	
	B3LYP DFT-D	−7.79	−6.81	
	WB97XD	−7.42	−6.51	
	M06	−7.00	−6.15	
	M062X	−7.59	−6.63	
TMG + 1 water complex (WD)	MP2	−7.46	−4.93	3
	HF	−4.71	−3.88	
	B3LYP	−5.38	−4.58	
	B3LYP DFT-D	−8.92	−7.84	
	WB97XD	−8.01	−7.01	
	M06	−7.39	−6.46	
	M062X	−8.48	−7.37	
TMG + 2 water complex	MP2	−16.39	−11.30	4
	HF	−10.80	−9.24	
	B3LYP	−13.46	−11.44	
	B3LYP DFT-D	−18.32	−16.10	
	WB97XD	−16.91	−14.81	
	M06	−16.33	−14.19	
	M062X	−17.90	−15.51	
TMG dimer	MP2	−9.46	−5.51	4
	HF	−4.52	−3.51	
	B3LYP	−7.58	−6.61	
	B3LYP-DFTD	−13.38	−12.26	
	WB97XD	−9.96	−8.90	
	M06	−14.69	−13.68	
	M062X	−15.18	−13.87	
Water dimer	MP2	−6.39	−4.83	1
	HF	−5.01	−4.36	
	B3LYP	−5.97	−5.19	
	B3LYP-DFTD	−6.71	−5.95	
	WB97XD	−6.35	−5.61	
	M06	−5.99	−5.25	
	M062X	−6.58	−5.80	

The potential energy curves for a free O–H (O12–H13) bond of single TMG molecule and hydrogen-bonded O–H (O12–H13) of TMG + 1 water complex (TD) are presented in Figures 3(a) and 3(b), respectively. The broadening of potential energy curve and appearance of asymmetrical double minimum in potential energy curve of hydrogen-bonded O–H reveal that a moderately strong hydrogen bond (O14⋯H13) is formed between TMG and water molecule [2]. The interaction energy barrier is high, which provides allowances for having various energetically lower protonic states [9].

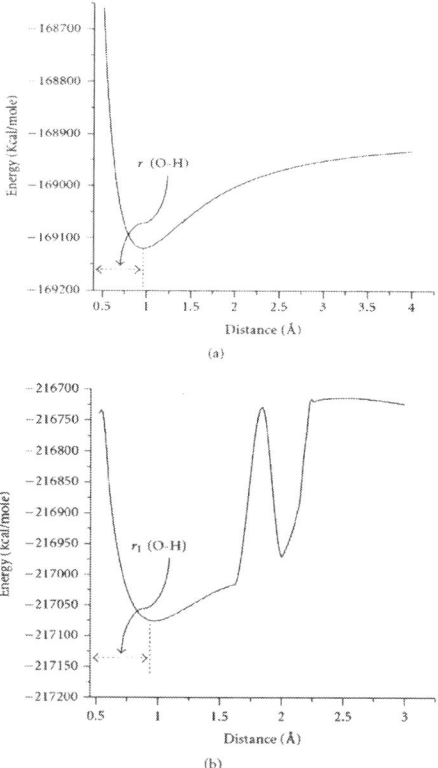

Figure 3: Calculated energy (kcal/mol) curve using WB97XD/6-31++G(d,p) for (a) a free bond of O–H (O12–H13) group of single TMG molecule (refer Figure 1(b)) and (b) hydrogen-bonded O–H (O12–H13) group of TMG + 1 water complex (TD) (refer Figure 2(b)).

Absolute Mullikan charge difference, absolute NBO charge difference, and absolute Chelpg charge difference between intermolecular hydrogen bond-forming oxygen and hydrogen atoms are obtained by taking absolute values of the difference between charge of oxygen and charge of hydrogen atoms and summarized in Table 3. As absolute charge differences between two atoms increase, the attractive electrostatic force between those two atoms also increases. Calculated absolute Mullikan charge difference, absolute NBO charge difference, and absolute Chelpg charge difference between intermolecular hydrogen bond-forming atoms (O14, H13) for TMG + 1 water complex (TD) are maximum and accordingly, forming strongest hydrogen bond compared to other systems for all the methods used in this paper, as shown in Table 3.

Table 3: Calculated absolute Mullikan charge difference, absolute NBO charge difference, and absolute Chelpg charge difference between hydrogen bond-forming atoms for TMG + n water complex (n=1,2) using 6-31++G(d,p) basis set and various methods

System	Methods	Hydrogen bond-forming atoms	Mullikan charge diff. (a.u.)	NBO charge diff. (a.u.)	Chelpg charge diff. (a.u.)
TMG + 1 water complex (TD)	MP2	O14···H13	1.25	1.54	1.39
	HF	O14···H13	1.23	1.53	1.43
	B3LYP	O14···H13	1.20	1.51	1.33
	B3LYP DFT-D	O14···H13	1.20	1.51	1.30
	WB97XD	O14···H13	1.20	1.52	1.33
	MO6	O14···H13	1.21	1.54	1.36
	MO62X	O14···H13	1.25	1.54	1.36

TMG + 1 water complex (WD)	MP2	O4···H16	1.02	1.34	1.14
		O12···H14	1.11	1.36	1.08
	HF	O4···H16	1.00	1.32	1.17
		O12···H14	1.09	1.34	1.11
	B3LYP	O4···H16	0.90	1.28	1.09
		O12···H14	1.01	1.30	1.03
	B3LYP DFT-D	O4···H16	0.91	1.28	1.09
		O12···H14	1.02	1.30	1.02
	WB97XD	O4···H16	0.93	1.29	1.10
		O12···H14	1.05	1.30	1.03
	MO6	O4···H16	0.94	1.30	1.09
		O12···H14	1.06	1.32	1.03
	MO62X	O4···H16	0.97	1.30	1.09
		O12···H14	1.08	1.31	1.01
TMG + 2 water complex	MP2	O4···H19	1.06	1.34	1.17
		O12···H16	1.17	1.38	0.93
	HF	O4···H19	1.02	1.33	1.19
		O12···H16	1.14	1.36	0.97
	B3LYP	O4···H19	0.93	1.28	1.13
		O12···H16	1.06	1.32	0.85
	B3LYP DFT-D	O4···H19	0.94	1.28	1.13
		O12···H16	1.09	1.32	0.87
	WB97XD	O4···H19	0.96	1.29	1.13
		O12···H16	1.10	1.32	0.89
	MO6	O4···H19	0.99	1.31	1.00
		O12···H16	1.13	1.34	0.80
	MO62X	O4···H19	1.01	1.30	1.00
		O12···H16	1.15	1.34	0.79

Highest occupied molecular orbital (HOMO) and lowest unoccupied molecular orbital (LUMO) of TMG and water systems, simulated by WB97XD/6-31++G(d,p) method, are presented in Figure 4. The LUMO energies of TMG + n water complexes (n=1,2) are less compared to that of single TMG and water molecule. The LUMO of TMG + 1 water complex (TD) originates essentially from the LUMO of water with negligible contribution of antibonding orbital of TMG, but the HOMO of the same complex arises largely from the HOMO of TMG. On the other hand, for TMG + 1 water complex (WD) and TMG + 2 water complex, LUMO comes mainly from the LUMO of the TMG, and HOMO is from the intermixing of lone pairs of both TMG and water molecules. Intermolecular hydrogen bond (O14⋯H13) of TMG + 1 water complex (TD) has very high covalent character compared to two intermolecular hydrogen bonds (O12⋯H14, O4⋯H16) of TMG + 1 water complex (WD). It is also justified by the respective hydrogen bond lengths, that is, 1.87 Å, 2.15 Å, and 2.18 Å for O14⋯H13, O12⋯H14, and O4⋯H16, respectively, as evident in Figures 2(b) and 2(c). In case of TMG + 2 water complex, the HOMO originates from major intermixing of lone pairs of TMG molecule and one water molecule (H19–O17–H18) and hardly any contribution from other water molecule (H16–O14–H15). It is also found from Figure 4, that the LUMO for TMG + 2 water complex originates from major intermixing of antibonding orbital of TMG molecule and one water molecule (H19–O17–H18). The covalent character is more prominent in one intermolecular hydrogen bond (O12⋯H16) compared to other hydrogen bond (O4⋯H19) in TMG + 2 water complex, and consequently the O12⋯H16 hydrogen bond is comparatively more strong, which is also supported by their hydrogen bond distances shown in Figure 1(d). Mixing of the HOMO of proton donor (O12–H13-bonding orbital of TMG) with the LUMO of proton acceptor (O14 of water molecule) in TMG + 1 water complex (TD) leads to decrease of electron density around O12–H13 bond.

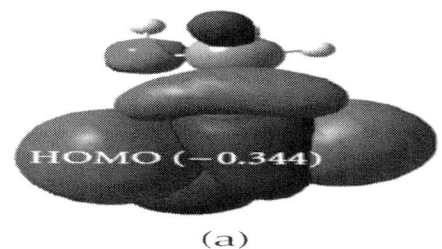

(a)

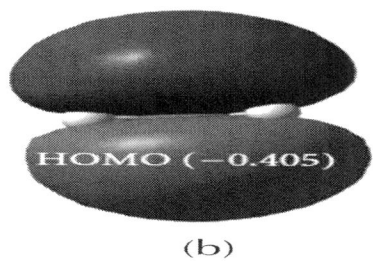

(b)

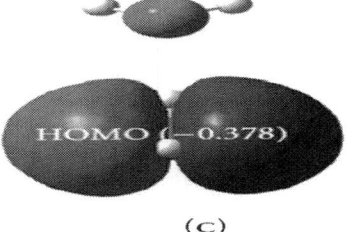

(c)

(d)

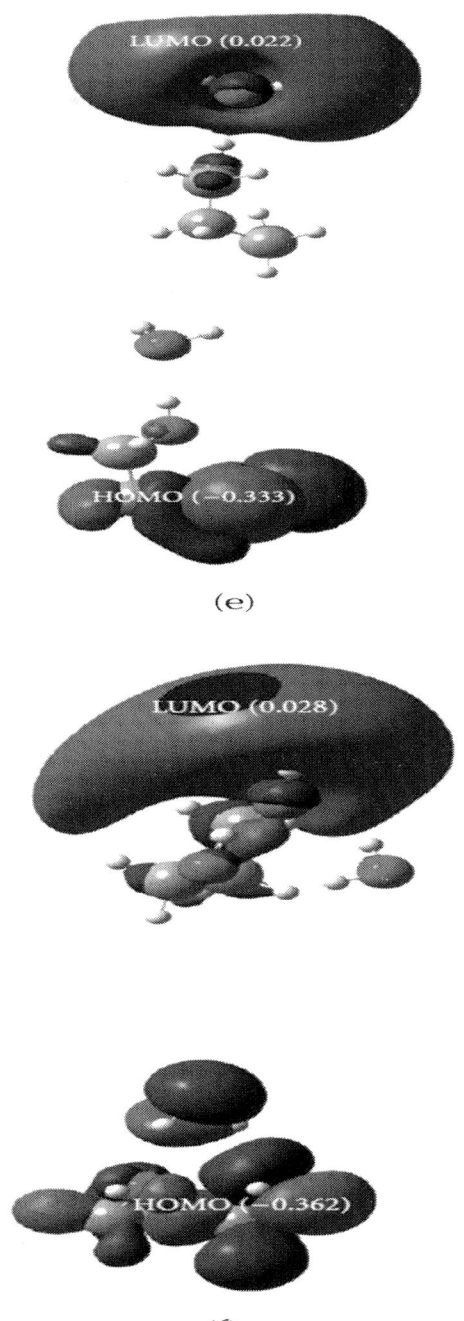

(e)

(f)

(g)

Figure 4: Frontier orbitals (HOMO, LUMO energies is atomic unit) of (a) TMG monomer, (b) water monomer, (c) water dimer, (d) TMG dimer, (e) TMG + 1 water complex (TD), (f) TMG + 1 water complex (WD), and (g) TMG + 2 water complex by WB97XD/6-31++G(d,p) theory.

The calculated second-order perturbation energies and respective occupancies for selective donor-acceptor interactions relevant to hydrogen bond formation in single TMG molecule and TMG + 1 water (TD) complex from NBO analysis are given in Table 4. Calculated second-order perturbation energy of donor- (lone pair of O14) acceptor (antibonding orbital of O12–H13) interaction responsible for intermolecular hydrogen bonding is higher than that of donor- (lone pair of O12) acceptor (antibonding orbital of O4–H5) interaction responsible for intramolecular hydrogen bonding in TMG + 1 water complex (TD) according to all the methods used in this work. It is inferred that the intermolecular hydrogen bond is stronger than intra molecular hydrogen bond for TMG-water complex (TD) as supported by respective hydrogen bond distances.

Table 4: Calculated second-order perturbation energy ($\Delta E_{ij}^{(2)}$, kcal/mol) for TMG single and TMG − n water complex (n=1,2) using 6-31++G(d,p) basis set and various methods

System	Method	Donor	Occu-pancyof donor	Acceptor	Occu-pancyof acceptor	$\Delta E_{ij}^{(2)}$
TMG	MP2	LP(2)O12	1.973	BD*(1)O4–H5	0.013	4.88 (Intra)
	HF	LP(1)O12	1.984	BD*(1)O4–H5	0.011	3.07 (Intra)
	B3LYP	LP(2)O12	1.979	BD*(1)O4–H5	0.019	4.72 (Intra)
	B3LYP DFT-D	LP(2)O12	1.962	BD*(1)O4–H5	0.018	4.17 (Intra)
	WB97XD	LP(2)O12	1.963	BD*(1)O4–H5	0.018	5.09 (Intra)
	MO6	LP(2)O12	1.961	BD*(1)O4–H5	0.018	3.20 (Intra)
	MO62X	LP(2)O12	1.965	BD*(1)O4–H5	0.016	3.63 (Intra)
TMG + 1 water complex (TD)	MP2	LP(2)O12	1.968	BD*(1)O4–H5	0.017	7.28 (Intra)
		LP(2)O14	1.981	BD*(1)O12–H13	0.020	15.79 (Inter)
	HF	LP(1)O12	1.981	BD*(1)O4–H5	0.012	3.74 (Intra)
		LP(2)O14	1.988	BD*(1)O12–H13	0.014	10.12 (Inter)
	B3LYP	LP(2)O12	1.955	BD*(1)O4–H5	0.023	6.03 (Intra)
		LP(2)O4	1.978	BD*(1)O12–H13	0.030	14.30 (Inter)
	B3LYP DFT-D	LP(2)O12	1.952	BD*(1)O4–H5	0.025	7.07 (Intra)
		LP(2)O14	1.969	BD*(1)O12–H13	0.033	16.57 (Inter)
	WB97XD	LP(2)O12	1.956	BD*(1)O4–H	0.022	7.42 (Intra)
		LP(2)O14	1.972	BD*(1)O12–H13	0.030	18.04 (Inter)
	MO6	LP(2)O12	1.953	BD*(1)O4–H5	0.022	5.56 (Intra)
		LP(2)O14	1.976	BD*(1)O12–H13	0.026	13.09 (Inter)
	MO62X	LP(2)O12	1.960	BD*(1)O4–H5	0.019	5.05 (Intra)
		LP(2)O14	1.977	BD*(1)O12–H13	0.025	14.68 (Inter)

The charge transfer (CT) energies calculated using natural energy decomposition analysis (NEDA) for TMG, TMG + 1 water complex (TD), TMG + 1 water complex (WD), TMG + 2 water complex, water dimer, and TMG dimer are presented as bar chart in Figure 5. Charge transfer (CT) is a part of the stabilization energy of intermolecular interacting system [46], and it plays an important role in hydrogen bond formation [47, 48]. CT represents electron delocalization interaction between occupied molecular orbital of one molecule and unoccupied molecular orbital of another molecule. As TMG dimer shows lowest CT value compared to that of others, TMG dimer would be having the strongest intermolecular hydrogen bond interaction. Consequently, TMG would be effective for inhibiting water cluster formation only when there is no favorable TMG dimer formation condition, that is, at low concentration of TMG.

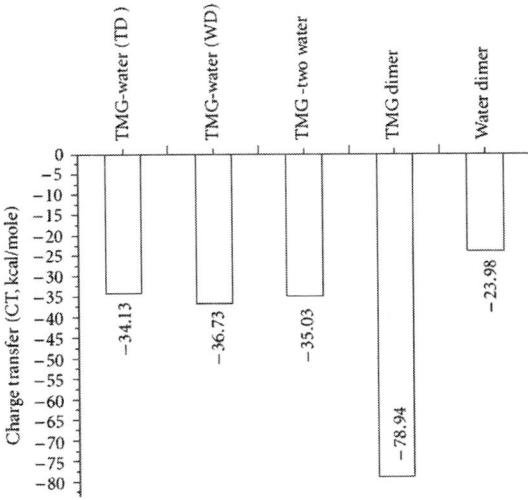

Figure 5: Bar chart of calculated charge transfer (CT, kcal/mol) by WB97XD/6-31++G(d,p) theory.

Calculated electron density contours with hydrogen bond critical point (HBCP), total electron density, and total Laplacian electron density at HBCP using AIM analysis for TMG molecule, water dimer, and TMG + 1 water complex (TD) using WB97XD/6-31++G(d,p) method are represented in Figure 6 and Table 5, respectively. One hydrogen bond critical point is determined for

TMG molecule and water dimer, but two hydrogen bond critical points are found in TMG + 1 water complex (TD). The hydrogen bond critical point (HBCP) is a specific point between the donor and acceptor, where the gradient of electron density is zero, and it is essential evidence of hydrogen bond existence. TMG + 1 water complex (TD) has more covalence character and consequently more strength compared to water dimer as it has higher electron density at HBCP [41, 49] as evident in Table 5.

Table 5: Calculated total electron density $\left(\sum \rho(r_c),(e/a^3)\right)$, total Laplacian electron density $\left(\Sigma \nabla^2 \rho(r_c),(e/a^5)\right)$ for TMG molecule, water dimer, and TMG + 1 water (TD) complex at hydrogen bond critical point (HBCP) using WB97XD/6-31++G(d,p)

Systems	$\sum \rho(r_c)$	$\sum \nabla^2 \rho(r_c)$
Single TMG	0.0220	0.0701
Water dimer	0.0228	0.0645
TMG + 1 water complex (TD)	0.0291	0.0771

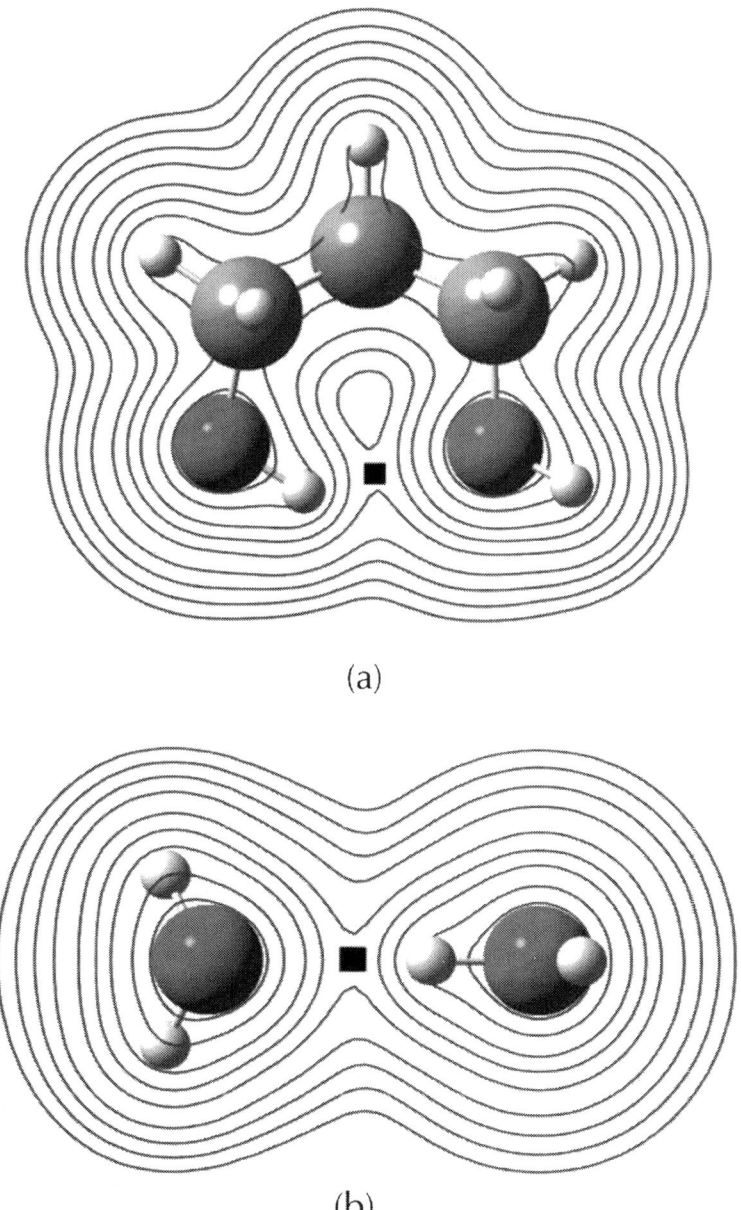

(a)

(b)

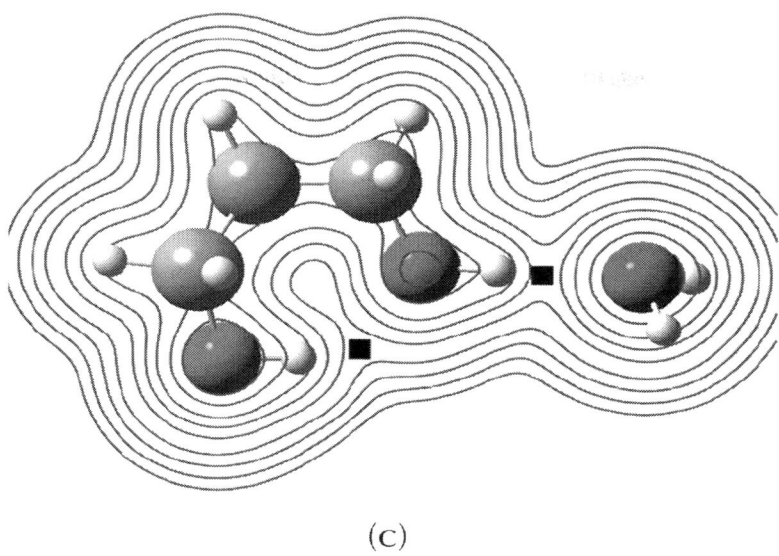

(c)

Figure 6: Contour map of the electron density for (a) single TMG molecule, (b) water dimer, (c) TMG + 1 water complex (TD) by WB97XD/6-31++G(d,p) theory. Hydrogen bond critical points are indicated by filled square symbol, ■ (colour legend: red = oxygen, black = carbon and whitish grey = hydrogen).

Calculated vibrational frequencies and IR intensities of O–H stretching of water and TMG molecule, TMG dimer, and TMG + n water complex system (n=1,2) using WB97XD/6-31++G(d,p) are listed in Table 6. The red shift and intensity of hydrogen-bonded O–H stretching for TMG dimer is higher than that of TMG + n water complex (n=1,2). It is also detected that the red shift and IR intensity of hydrogen-bonded O–H stretching of TMG molecule in TMG + 1 water (TD) system are higher than that of O–H stretching of water molecule in TMG + 1 water (WD) system. As higher values of red shift and intensity for hydrogen-bonded O–H bond stretching indicate stronger hydrogen bond, intermolecular hydrogen bond in TMG + 1 water (TD) system is stronger compared to the intermolecular hydrogen bond in TMG + 1 water (WD) system. Consequently, TMG has higher tendency to act as a proton donor to form hydrogen bond with water molecule.

Table 6: Calculated scaled vibrational frequency (cm^{-1}), red shift (cm^{-1}), IR intensity (km-mol^{-1}) of O–H bond stretching for water molecule, and TMG molecule and TMG + n water complex (n=1,2) using WB97XD/6-31++G(d,p) along with some experimental vibrational frequency (cm^{-1})

System	O–Hbondstretchingforwatermolecule				O–HbondstretchingforT-MGmolecule		
	Scaled-freq.	Red-shift	IRinten-sity	Exp.vibra-tionalfre-quency	Scaled-freq.	Red-shift	IRinten-sity
Watermol-ecule	3802		8.2	3756[9]			
TMGmolecule					3757		144.43
TMGdimer					3519	238	585.61
TMG+1water complex(TD)	3797	5	13.87		3663	94	717.93
TMG+1water complex(WD)	3736	66	95.97		3754	3	80.39
TMG+2water complex	3665	137	512.89		3766	—	133.90

CONCLUSIONS

A thorough analysis of hydrogen bond formation in trimethylene glycol (TMG) + n water complex (n=1,2) has been performed based on calculated interaction energies, NBO, AIM, charge transfer, and red shift using HF, MP2, DFT, and DFT-D methods. TMG + 2 water complex, found to be most stable compared to TMG + 1 water complexes, TMG dimer, and water dimer as per calculated interaction energies. For TMG + 1 water complex, stronger intermolecular hydrogen bond formed when TMG acts as a proton donor as per charge differences between respective hydrogen bond-forming atoms, NBO analysis, and red shifts of calculated vibrational spectra. The broadening as well as asymmetrical double minimum appearance in potential energy curve of hydrogen-bonded O–H reveals that a moderately strong hydrogen bond (O14···H13) is formed in TMG + 1 water complex (TD). Intermolecular hydrogen bond of TMG + 1 water complex (TD) has higher covalent character and accordingly, higher strength compared to that of TMG + 1 water complex (WD) as per HOMO-LUMO study. The hydrogen bond in TMG dimer is found to be stronger compared to other systems as per calculated charge transfer and red shift values. Very dilute TMG solution is recommended in order to break water cluster. This work illustrates electronic structure property correlation-based understandings of trimethylene glycol in aqueous solution and would help in designing inhibitors for water cluster/clathrate system like methane hydrate.

ACKNOWLEDGMENTS

This work is financially supported by Ministry of Earth Science, Government of India (Project no. MoES/16/48/09—RDEAS (MRDM5)). The authors also acknowledge Accelrys Inc. for providing free Discovery studio 3.1 visualization tool.

REFERENCES

1.   P. Schuster and P. Wolschann, "Hydrogen bonding: from small clusters to biopolymers," Monatshefte fur Chemie, vol. 130, no. 8, pp. 947–960, 1999.

2.   G. A. Jeffrey, An Introduction to Hydrogen Bonding, Oxford University Press, New York, NY, USA, 1997.

3.   A. Demirbas, Methane Gas Hydrate, Springer, London, UK, 2010.

4.   T. S. Collett, "Energy resource potential of natural gas hydrates," AAPG Bulletin, vol. 86, no. 11, pp. 1971–1992, 2002.

5.   P. Englezos, "Clathrate hydrates," Industrial and Engineering Chemistry Research, vol. 32, no. 7, pp. 1251–1274, 1993.

6.   E. G. Hammerscht, "Formation of gas hydrates in natural gas transmission lines," Industrial & Engineering Chemistry Research, vol. 26, no. 8, pp. 851–855, 1984.

7.   J. K. Fink, Petroleum Engineer's Guide to Oil Field Chemicals and Fluids, Elsevier, Oxford, UK, 2012.

8.   A. Wehner, R. Miller, G. Fenyvesi, J. W. DeSalvo, and M. Joerger, "Heat transfer compositions comprising renewable-based biodegradable 1, 3-propanediol," US patent 2007/0200088 A1, 2007.

9.   V. May and O. Kühn, Charge and Energy Transfer Dynamics in Molecular Systems, Wiley-VCH, Weinheim, Germany, 2005.

10.   S. J. Grabowski, T. L. Robinson, and J. Leszczynski, "Strong dihydrogen bonds—Ab initio and atoms in molecules study," Chemical Physics Letters, vol. 386, no. 1–3, pp. 44–48, 2004.

11.   S. Wojtulewski and S. J. Grabowski, "DFT and AIM studies on two-ring resonance assisted hydrogen bonds," Journal of Molecular Structure, vol. 621, no. 3, pp. 285–291, 2003.

12.   S. Pal and T. K. Kundu, "Dodecahedron methane hydrate cage structure—an Ab initio study," Journal of Petroleum Engineering and Technology, vol. 2, pp. 22–35, 2012.

13.   D. Peeters, "Hydrogen bonds in small water clusters: a theoretical point of view," Journal of Molecular Liquids, vol. 67, pp. 49–61, 1995.

14.   X. M. Zhou, Z. Y. Zhou, H. Fu, Y. Shi, and H. Zhang, "Density functional complete study of hydrogen bonding between the dichlorine monoxide and the hydroxyl radical (Cl2O·HO)," Journal of Molecular Structure, vol. 714, no. 1, pp. 7–12, 2005.

15.   P. K. Sahu, A. Chaudhari, and S. L. Lee, "Theoretical investigation for the hydrogen bond interaction in THF-water complex," Chemical Physics Letters, vol. 386, no. 4–6, pp. 351–355, 2004.

16.   P. K. Sahu and S. L. Lee, "Hydrogen-bond interaction in 1:1 complexes of tetrahydrofuran with water, hydrogen fluoride,

and ammonia: a theoretical study," Journal of Chemical Physics, vol. 123, no. 4, Article ID 044308, 9 pages, 2005.

17. A. Mandal, M. Prakash, R. M. Kumar, R. Parthasarathi, and V. Subramanian, "Ab Initio and DFT studies on methanol-water clusters," Journal of Physical Chemistry A, vol. 114, no. 6, pp. 2250–2258, 2010.

18. J. E. Del Bene, "An ab initio study of the structures and enthalpies of the hydrogen-bonded complexes of the acids H_2O, H_2S, HCN, and HCl with the anions OH-, SH-, CN-, and Cl-," Structural Chemistry, vol. 1, no. 1, pp. 19–27, 1990.

19. I. Alkorta, F. Blanco, P. M. Deyà et al., "Cooperativity in multiple unusual weak bonds," Theoretical Chemistry Accounts, vol. 126, no. 1, pp. 1–14, 2010.

20. I. Mata, E. Molins, I. Alkorta, and E. Espinosa, "Topological properties of the electrostatic potential in weak and moderate N···H hydrogen bonds," Journal of Physical Chemistry A, vol. 111, no. 28, pp. 6425–6433, 2007.

21. J. B. Levy, N. H. Martin, I. Hargittai, and M. Hargittai, "Intra- and intermolecular hydrogen bonding in 2-phosphinylphenol: a quantum chemical study," Journal of Physical Chemistry A, vol. 102, no. 1, pp. 274–279, 1998.

22. O. V. Shishkin, I. S. Konovalova, L. Gorb, and J. Leszczynski, "Novel type of mixed O-H···N/O-H···πhydrogen bonds: monohydrate of pyridine," Structural Chemistry, vol. 20, no. 1, pp. 37–41, 2009.

23. V. Horváth, A. Kovács, and I. Hargittai, "Structural aspects of donor-acceptor interactions," Journal of Physical Chemistry A, vol. 107, no. 8, pp. 1197–1202, 2003. ·

24. C. C. J. Roothaan, "New developments in molecular orbital theory," Reviews of Modern Physics, vol. 23, no. 2, pp. 69–89, 1951.

25. M. Head-Gordon, J. A. Pople, and M. J. Frisch, "MP2 energy evaluation by direct methods," Chemical Physics Letters, vol. 153, no. 6, pp. 503–506, 1988.

26. P. Hohenberg and W. Kohn, "Inhomogeneous electron gas," Physical Review, vol. 136, no. 3, pp. B864–B871, 1964.

27. W. Kohn and L. J. Sham, "Self-consistent equations including exchange and correlation effects,"Physical Review, vol. 140, no. 4, pp. A1133–A1138, 1965. ·

28. S. Grimme, "Accurate description of van der Waals complexes by density functional theory including empirical corrections," Journal of Computational Chemistry, vol. 25, no. 12, pp. 1463–1473, 2004.

29. A. D. Becke, "Density-functional exchange-energy approximation with correct asymptotic behavior,"Physical Review A, vol. 38, no. 6, pp. 3098–3100, 1988. ·

30. C. Lee, W. Yang, and R. G. Parr, "Development of the Colle-Salvetti correlation-energy formula into a functional of the electron density," Physical Review B, vol. 37, no. 2, pp. 785–789, 1988.

31. J. D. Chai and M. Head-Gordon, "Long-range corrected hybrid density functionals with damped atom-atom dispersion corrections," Physical Chemistry Chemical Physics, vol. 10, no. 44, pp. 6615–6620, 2008.

32. Y. Zhao and D. G. Truhlar, "The M06 suite of density functionals for main group thermochemistry, thermochemical kinetics, noncovalent interactions, excited states, and transition elements: two new functionals and systematic testing of four M06-class functionals and 12 other functionals," Theoretical Chemistry Accounts, vol. 120, no. 1–3, pp. 215–241, 2008. ·

33. P. C. Hariharan and J. A. Pople, "The influence of polarization functions on molecular orbital hydrogenation energies," Theoretica Chimica Acta, vol. 28, no. 3, pp. 213–222, 1973.

34. J. Chandrasekhar, J. G. Andrade, and P. Von Ragué Schleyer, "Efficient and accurate calculation of anion proton affinities," Journal of the American Chemical Society, vol. 103, no. 18, pp. 5609–5612, 1981.

35. M. S. Gordon and J. H. Jensen, "Understanding the hydrogen bond using quantum chemistry,"Accounts of Chemical Research, vol. 29, no. 11, pp. 536–543, 1996.

36. S. F. Boys and F. Bernardi, "The calculation of small molecular interactions by the differences of separate total energies. Some procedures with reduced errors," Molecular Physics, vol. 19, no. 4, pp. 553–566, 1970.

37. F. Weinhold and C. R. Landis, "Natural bond orbitals and extensions of localized bonding concepts,"Chemistry Education Research and Practice, vol. 2, pp. 91–104, 2001.

38. E. D. Gledening, A. E. Reed, J. A. Carpenter, and F. Weinhold, NBO. version 3.1.

39. A. E. Reed, L. A. Curtiss, and F. Weinhold, "Intermolecular interactions from a natural bond orbital, donor-acceptor viewpoint," Chemical Reviews, vol. 88, no. 6, pp. 899–926, 1988.

40. A. Y. Li, "Chemical origin of blue- and red shifted hydrogen bonds: intra-molecular hyper-conjugation and its coupling with intermolecular hyper-conjugation," Journal of Chemical Physics, vol. 126, pp. 154102–154111, 2007.

41. R. F. W. Bader, "Atoms in molecules," Accounts of Chemical Research, vol. 18, pp. 9–15, 1985.

42. M. J. Frisch, G. W. Trucks, H. B. Schlegel, et al., "Gaussian 09, Revision (B.01)," Gaussian Inc., Wallingford CT, 2010.

43. I. M. Alecu, J. Zheng, Y. Zhao, and D. G. Truhlar, "Computational thermochemistry: scale factor databases and scale factors for vibrational frequencies obtained from electronic model chemistries,"Journal of Chemical Theory and Computation, vol. 6, no. 9, pp. 2872–2887, 2010. ·

44. B. Civalleri, C. M. Zicovich-Wilson, L. Valenzano, and P. Ugliengo, "B3LYP augmented with an empirical dispersion term (B3LYP-D*) as applied to molecular crystals," CrystEngComm, vol. 10, no. 4, pp. 405–410, 2008.

45. A. E. Lutskii and N. I. Gorokhova, "Intramolecular hydrogen bonds and molecular dipole moments,"Theoretical and Experimental Chemistry, vol. 4, no. 6, pp. 532–534, 1971.

46. H. Umeyama and K. Morokuma, "Origin of alkyl substituent effect in the proton affinity of amines, alcohols, and ethers," Journal of the American Chemical Society, vol. 98, no. 15, pp. 4400–4404, 1976.

47. H. Umeyama and K. Morokuma, "The origin of hydrogen bonding. An energy decomposition study,"Journal of the American Chemical Society, vol. 99, no. 5, pp. 1316–1332, 1977.

48. A. Van der Vaart and K. M. Merz Jr., "Charge transfer in small hydrogen bonded clusters," Journal of Chemical Physics, vol. 116, no. 17, pp. 7380–7388, 2002.

49. S. J. Grabowski, Hydrogen Bonding-New Insights, Springer, Dordrecht, The Netherlands, 2006.

Chapter 2

Improvement of Natural Gas Dehydration Performance by Optimization of Operating Conditions: A Case Study in Sarkhun Gas Processing Plant

M.R. Rahimpour[a, b], M. Saidi[a], and M. Seifi[a, c]

[a]Department of Chemical Engineering, School of Chemical and Petroleum Engineering, Shiraz University, Shiraz 71345, Iran
[b]Gas Center Of Excellence, Shiraz University, Shiraz 71345, Iran
[c]N.I.G.C Sarkhun & Qeshm Gas Treating Co., Bandar Abbas, Iran

ABSTRACT

Water dew point adjustment unit is one of the main natural gas refinement processes for controlling the content of water and other impurities in natural gas to some allowable limit. Investigating the dew point adjustment unit of the Sarkhun gas processing facility revealed that due to inappropriate performance of liquid level control of the separators, this unit encountered hydrate formation and also the glycol consumption rate increased considerably. In this regard, the three phase separator and filters were simulated by steady state simulation software and the optimum separation temperature was determined. The operating condition of the dew point control unit was adjusted based on the optimum conditions achieved by simulation and for more investigation, experimental sampling was performed. The experimental results showed that the water and hydrocarbon dew point temperatures improved from $-10°C$ to $-26°C$ and from $-5°C$ to $-9°C$, respectively after modifications. The outstanding achievement was a remarkable reduction in greenhouse gas emission after modifications mainly due to a considerable decrease in gas flared. According to experimental data, greenhouse gas production reduced by about 50% (decreased from 60,000 to 30,000 ton/year). More importantly, greater than 6000 tonnes of Liquified Petroleum Gas (LPG) was additionally produced annually per 1 MMSCM that provided additional annual income of about $11 million.

INTRODUCTION

One of the main processes in natural gas treatment is elimination of impurities such as water, heavy hydrocarbons, hydrogen sulfide, N_2, CO_2, and etc. from the wellhead gas stream. In the gas refineries, according to the type of the feed characteristics and the inlet molar flow rates, several treatment units are established (Kohl and Nielsen, 1997 and Katz, 1959). Dew point adjustment is a key natural gas treatment process for decreasing the content of water and other impurities to some allowable value (Rojey and Jaffret, 1997 and Faulkner, 2006). Due to the occurrence of some feed changes, equipment limitations and poor operation such as changing in feed characteristics, operating conditions, degradation of facilities, chemical contaminants and environmental considerations, process optimization is necessary to improve the performance of process plants (Rahimpour et al., 2011). Gas hydrate formation due to the presence of water in the gas stream is one the most important limitations in the natural gas treatment and transition industry. In the gas dehydration unit, water separation from the gas stream via dew point adjustment not only prevents hydrate formation and pipeline blockage, but also decreases the probability of equipment corrosion (Mokhatab and Poe, 2006).

Løkken et al. (2008) studied different methods of water dew point control and investigated general phase behavior of natural gas with water and other trace components such as glycols. Also they developed a thermodynamic model for prediction of equilibrium water vapor concentration and natural gas dew point. In another related work, Erdmann et al. considered the natural gas dew point adjustment and performed a parametric sensitivity analysis (Ale et al., 2009).

Another major concern of the natural gas industry is emissions of benzene, toluene, ethylbenzene, and xylene isomers (BTEX) and other volatile organic compounds (VOC) which vent from the reboiler of glycol dehydration units (Rueter et al., 1993). Many researchers worked on prevention of BTEX, VOC and greenhouse gases emissions to atmosphere. Peinemann and Stegger developed a gas permeation membrane which was applicable in hydrocarbon recovery processes such as BTEX and fluorochlorinated hydrocarbons, propylene recovery process and gas conditioning processes such as water vapor-dew point adjustment, methane number enhancement, natural gas drying (Peinemann and Stegger, 2003). Darwish and Hilal simulated a typical natural gas dehydration process to study the effects of different input parameters and operating conditions of the absorption column, the stripper and the overall plant on VOC emissions and water content of the dehumidified natural gas (Darwish and Hilal, 2008). BTEX and other VOC emissions are environmental challenges for the natural gas industry. A pressurized glycol reboiler is an effective factor in controlling VOC emissions from glycol gas-dehydration systems (Hicks et al., 2004). The application of deliquescing desiccants (Hygroscopic salts such as calcium chloride, lithium chloride and potassium chloride) offers many advantages over traditional drying methods such as triethylene glycol, including operational simplicity, no VOC emissions, low capital and maintenance costs (Acor and Mirdadian, 2003). One method of reducing BTEX emissions is the use of the proper dehydrating agent. BTEX compounds are less soluble in diethylene glycol (DEG) than triethylene glycol (TEG) and considerably less soluble in ethylene glycol (EG) (Covington et al., 1998). An effective economic method to reduce emissions of VOC is burning a mixture of condensable and non-condensable gases as a fuel in the reboiler (Grizzle, 1993). Branco et al. (2010) evaluated the possibility of installing an offshore gas-to-liquids (GTL) plant to reduce natural gas flaring and control carbon dioxide equivalent emissions. In another related study, Conoco has developed a glycol dehydrator process enhancement to recover hydrocarbons normally vented with water vapor during glycol regeneration (Choi and Spisak, 1993). Khosravanipour et al. (2009) investigated thermodynamic behavior of water + triethylene glycol (TEG) and water + TEG + toluene systems. They reported that in these systems, the volatility of water in water + TEG solutions enhanced considerably with the higher purity of dehydrated TEG achieved. This was also enhanced by the addition of toluene.

There are several glycol regeneration processes, each applies different methods to enhance process performance by reducing the partial pressure of water in the vapor phase (Twu et al., 2005). Pearce et al. considered the Drizo process in which gas dehydration is performed using glycol as the dehydrating agent and toluene as a regeneration solvent (Pearce et al., 1972). Paymooni et al. (2011) proposed addition of liquid hydrocarbon solvents such as toluene and isooctane to enhance the glycol regeneration process. Their results indicated that the liquid hydrocarbon solvent addition can considerably enhance glycol purity, water volatility and remarkably reduce glycol loss. These improvements are due to the fact the liquid hydrocarbon solvent vaporized rapidly in the reboiler and enhanced the water volatility, and as a result, it increased glycol concentration in the reboiler. In any operational gas delivery contract, the water and hydrocarbon dew points should be mutually acceptable. In 2007, some of the consumers complained about the presence of detergent like substances in the gas, therefore the Sarkhun gas plant was thoroughly investigated. The main issue with the dehydration unit was the increasing pressure drop of the exchangers and predominately the outlet filters that created unit control problems. Due to variability of the inlet gas feed, improper gas cooling, insufficient glycol injection and no water separation from gas before cooling, hydrate formation took place in the outlet gas filters and baffles of the cold separator (S-402). Due to blockage of gas pathways through the baffles of S-402, the gas continued to carry over significant entrained glycol and liquid hydrocarbons. After preheating in the heat exchangers, glycol transferred to the gas phase and increased the hydrocarbon dew point considerably. Furthermore, in units with the part blocked baffles (S-401) or improper glycol injection, hydrate formation took place in the outlet gas filters and blocked the gas pathway and caused the water dew point adjustment unit to be taken out of service.

In this research, the unit was modeled using steady state simulation software and the hydrate formation temperature and optimized separation temperature were determined. Afterward, experimental sampling was performed under optimized conditions.

PROCESS DESCRIPTION

Water Dew Point Adjustment Unit

A schematic diagram of the water dew point adjustment unit is depicted in Fig. 1. The saturated water content of natural gas is reduced with decreasing temperature or increasing pressure. Therefore, saturated hot gases could be partially dehydrated by decreasing temperature. This could be performed by using direct cooling of gas by heat exchanger or the Joule- Thompson (J.T.) phenomenon (Campbell, 1992). In the cooling method, gas is cooled to the desired temperature and water and hydrocarbon dew points are simultaneously reduced. Gas is cooled due to heat transfer with a refrigerant and consequently water and heavy hydrocarbons were separated from the gas (Kohl and Nielsen, 1997 and Rojey and Jaffret, 1997). This process can be performed in several stages. This method is typically used for separating the majority of saturated water from gas streams (Rahimpour et al. 2013). Also, the gas temperature can be reduced below 0 °C by this method leading to more water separation from gas. A schematic diagram of dehydration by cooling method is depicted in Fig. 2(a). The gas to be dewpointed comes from the Slug Catcher Unit (Unit 200). As shown in Fig. 2 (a), after a decrease in temperature of separated gas from Unit 200 in gas–gas heat exchanger (E-401), it enters the Low Temperature Separator (S-401) where condensed liquid hydrocarbons and the water dissolved in glycol are separated in the subsequent separation stages. The operating pressure of separator (S-401) was almost 69 bar. The separated condensed liquid hydrocarbons and the obtained liquid hydrocarbons from the separation processes of Unit 200 are directed to water-liquid hydrocarbon separator (S-205). Details regarding S-205 can be found in Fig.2 (b). The separated liquid hydrocarbons are sent to the stabilization unit and the rich glycol routes to the Regeneration Unit (Unit 600). The process schematics of E-401, S-401 and S-205 are shown in Fig. 2 (b).

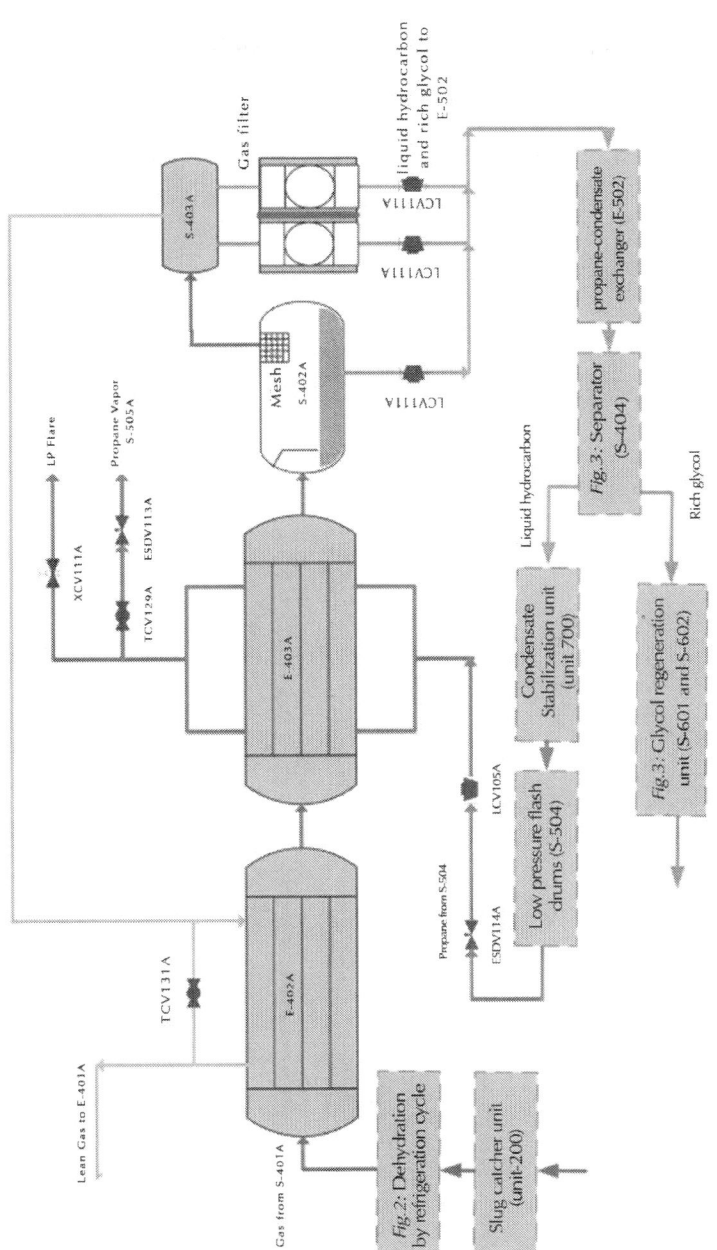

Figure 1: A schematic diagram of water dew point adjustment unit.

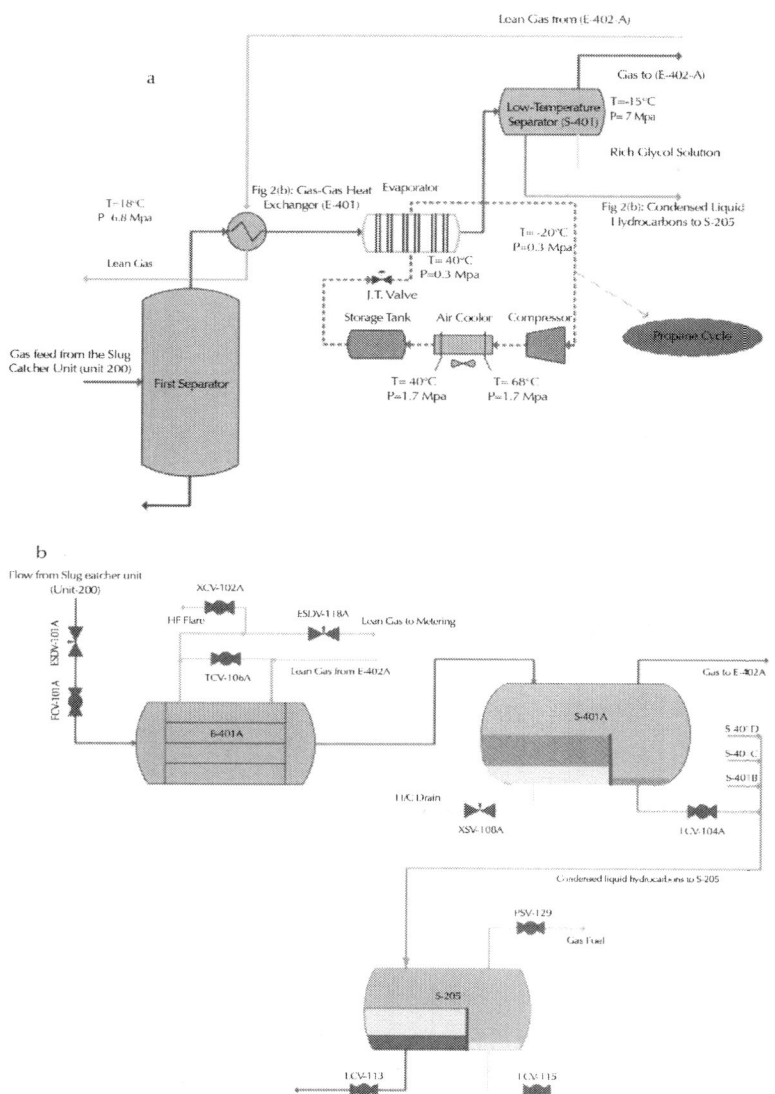

Figure 2: (a) A schematic diagram of dehydration by refrigeration cycle (b) process schematics of E-401, S-401 and S-205.

As shown in Fig. 2 (a), in the dehydration unit, the gas temperature is decreased in two stages by dry gas in gas–gas heat exchanger (E-401) and propane coolant cycle. Before these two stages, DEG is injected into the gas stream in order to prevent hydrate formation. Then as shown in Fig. 1, the separated gas from S-401 unit enters cold gas–gas heat exchanger (E-402). DEG is injected again to gas after leaving gas–gas heat exchanger, and then enters the chiller (E-403). The outlet fluid from chiller (E-403) consists of gas, liquid hydrocarbons and rich glycol. As depicted in Fig. 1, the remained liquid hydrocarbons and rich glycol are separated from gas in cold separator (S-402) with the operating pressure of 68.7 bar. The resulted gas enters final gas filter (S-403). All entrained liquid in gas is separated in outlet gas filter (S-403) which operates at pressure of 68.2 bar. The outlet dry gas from gas filter (S-403) is sent to gas transmission pipeline after flowing on the shell sides of gas–gas heat exchangers (E-402 and E-401). There are four dew point adjustment units (Unit 400 A-D) in the Sarkhun gas plant including water and hydrocarbon dew point adjustment units where each one has the capacity of 3.7 MMCM. These units are assisted with the purpose of absorbing the water vapor from Slug catcher (Unit 200) by DEG injection and also adjusting the hydrocarbon dew point of gas by separating liquid hydrocarbons and water and decreasing the related dew points.

Glycol Regeneration Unit

A schematic process diagram of regeneration unit is represented in Fig. 3. In this process, the separation of liquid hydrocarbons and glycol is done. The liquid outlet stream from the dew point adjustment unit (Fig. 1) is a mixture of glycol and liquid hydrocarbons that is sent to three phase separator (S-404). The mixture temperature is almost 30 °C which is sufficient for glycol separation from liquid hydrocarbons. Afterward, as shown in Fig. 3, rich glycol enters the liquid hydrocarbon/glycol separator (S-601) to separate the rich glycol from liquid hydrocarbons. This process is performed at 3 barg and 30 °C. The rich glycol exits from the bottom of the separator and send to the glycol concentrator. The outlet glycol stream is adjusted by level controller. Glycol temperature in reboiler reaches about 120 °C after passing through filters and being preheated by regenerated glycol. Thus, the water vapor is separated from rich glycol. The lean glycol, after heat exchanging with the inlet rich glycol and being cooled by air cooler, is send to its related vessels and pumped with 80 bar pressure to the dew point adjustment unit for injection to E-402 and E-403. Also the separated liquid hydrocarbon from the vessels is sent to the wet hydrocarbon drain section and pumped to the liquid stabilization unit after filtration.

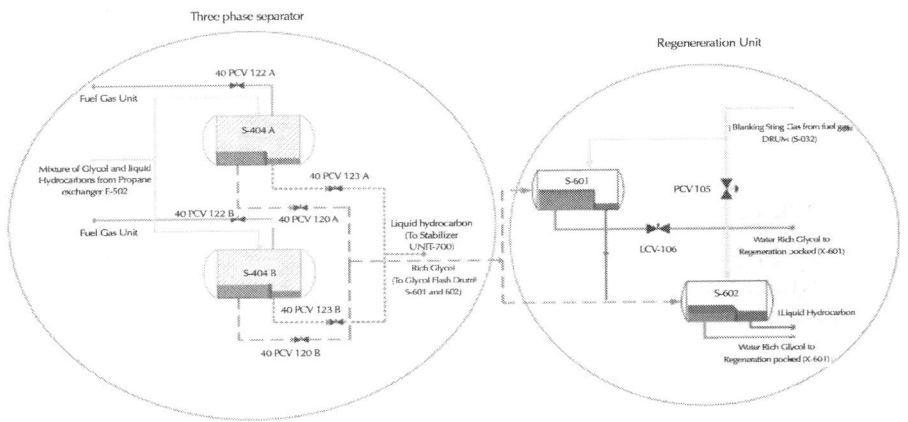

Figure 3: A schematic process diagram of three phase separator (S-404) and regeneration unit.

PROCESS CHALLENGES: A CASE STUDY IN SARKHUN GAS PROCESSING PLANT

Capacity Survey of the Glycol Circulation Rate

The Sarkhun gas plant investigations indicate that the reboiler coil and fourth heat exchanger caused 0.2 m³/hr decrease in flow rate.

Since the predominant pressure drop is in heat exchangers, it was recognized that the pressure of (S-601) should be increased from 3 barg to 5.5 barg. Thus, the flow rate of Unit 600 was increased from 3.4 m³/hr to 4 m³/hr. After several glycol cycle analyses, it was concluded that the glycol cycle had a good quality and only a small percent of it was oxidized.

The optimum separation temperature for liquid hydrocarbons from water was determined by simulation of a three phase separator. Fig. 3 shows a schematic diagram of three phase separator (S-404) for separation of liquid hydrocarbons and glycol. Investigating the operating condition of three phase separator (S-404) showed that its temperature should be maintained at 10 °C instead of 30 °C in most cases. Therefore, liquid hydrocarbons did not separate well from glycol. The liquid hydrocarbons went along with glycol to the regeneration unit. A tar-like liquid was produced owing to heating and cohesion of heavy molecules which caused further problems in the unit. Thus, controlling of separator temperature in the range of 25–30 °C was proposed.

Challenges of Dehydration Unit

The main problem of the dehydration unit was the increasing pressure drop in the heat exchangers, chillers and predominantly the outlet filters of dehydration unit (Unit 400). Unit controlling was disturbed when this problem took place. In order to maintain the outlet liquid hydrocarbon (H/C) in accordance with the contract with the customers, the outlet temperature from the chiller should be reduced which worsened the pressure drop problem in units. Unfortunately, liquids were not separated well in vessels upstream of the chiller. In slug catcher unit (S-200), entrained C_6^+ liquid spray did not separate well from the gas phase and it went along with the gas stream to the dew point adjustment unit. This problem was investigated in our previous work and addressed properly (Rahimpour et al. 2011). In order to minimize the mole percent of C_6^+ exited along with the outlet gas stream from the slug catcher, the pressure difference between the slug catchers and well streams were optimized (Rahimpour et al. 2011). The C_6^+ entrained percentage along with gas stream is presented in Table 1. Although the outlet gas stream from chiller was cool enough, the hydrocarbon dew point was not satisfied. A comparison between pressures drops in filters and hydrocarbon dew points are presented in Table 2. Considering the reported data of dew point adjustment unit of the Sarkhun gas processing facility represented that this unit encountered hydrate formation and high glycol consumption rate because of inappropriate performance of liquid level control of the separators. Thus this problem was investigated to find a proper solution. The three phase separator and filters were simulated and optimized to determine the appropriate temperature of separation process.

Table 1: The C_6^+ percentage entrained along with gas stream

Well no.	Separator no.	Mole % C_6^+ outlet (before adjustment)
1 & 15	S-101 O	1.96
2 &4	S-101 C	1.65
3&16	S-101 A	1.02
5	S-202 G	1.01
6	S-202 D	0.98
7	S-202 C	1.03
9	S-101 B	1.85
10	S-202 H	1.06
11	S-202 E	1.03
12	S-202 A	1.15
13	S-202 B	1.65
14	S-202 F	1.97

Table 2: A comparison between pressures drops in filters and hydrocarbon dew points

Date	H/C dew point (°C)	Water dew point (°C)	Gas from chiller Temp. (°C)	Final filter pressure drop (barg)
June 2007 at 13:13'	−3.6	−8.9	−13	0.77
July 2007 at 15:30'	−3.6	−8.7	−16	0.84
July 2007 at 7:35'	−5.7	−12.9	−15	0.34
July 2007 at 7:38'	−7.8	−15	−14.3	0.63

RESULTS AND DISCUSSION

Simulation Results

Decreasing the outlet temperature from the chiller increases the pressure drop and consequently the filters (S-403) were taken out of service to compensate for the pressure drop. As seen in Fig.4, the filter elements' conditions investigation indicated that the main reason for pressure drop was filters fouling via hydrate formation. As insufficient glycol injection would not prevent hydrate formation, total glycol injection rate to Unit 400 was evaluated and reported as 3.1 m³/h.

Figure 4: Hydrate formation in outlet filters and baffles of cold-separator (S-402) (Rahimpour et al., 2011).

Simulation of Dew Point Adjustment Unit and Hydrate Phenomenon Investigation

Since the gas in the tube side of (E-402) was not cooled enough, water vapor condensation decreased as well as the water elimination by glycol injection. Ultimately, the non separated vapor from gas in (E-402) entered the chiller and consequently hydrate formation took place as a result of insufficient amount of glycol injection. A comparison between hydrate formation temperatures in different ranges of glycol concentration is reported in Table 3. As indicated in this table and due to this fact that increasing the glycol concentration decreases the hydrate formation temperature, the glycol concentration was increased from 85% to 92%. The simulation results of hydrate formation temperature in chiller and heat exchangers of dehydration unit are depicted in Table 4 at two different glycol weight percent, 90 and 95%. The simulation results showed that the hydrate formation took place at much lower temperatures by increasing glycol concentration. The main reasons for hydrate formation were recognized as follows:

- Improper quality and quantity of injected glycol to gas as well as inappropriate location for glycol injection.

- Improper water and C_6^+ separation from the gas before entering chiller.

- Gas cooling to temperatures near to its hydrate formation limit

Table 3: A comparison between hydrate formation temperatures in different range of glycol concentration

Total glycol injection (m³/h)	Glycol to gas–gas exchanger (m³/h)	Glycol to chiller (m³/hr)	Glycol con. (%wt)	Hydrate temp. In gas–gas exchanger (°C)	Hydrate temp. In final filter (°C)	Chiller temp. (°C)
3.784	1.515	2.272	0.95	−10.63	−29.59	−12
3.784	0.756	3.030	0.95	−0.3	−29.59	−12
3.784	0.568	3.219	0.95	2.625	−29.59	−12

3.342	2.005	1.336	0.95	−15.9	−26.85	−12
3.342	1.336	2.005	0.95	−8.4	−26.82	−12
3.301	0.495	2.806	0.85	5.964	−12.44	−12
3.321	0.498	2.823	0.9	4.949	−18.5	−12

Table 4: The simulation results of hydrate formation temperature in chiller and exchangers for dehydration unit at 90 weight % and 95 weight % glycol

DEG composition (mass fraction)	90%	95%
Hydrate formation Temperature (°C)	−20.4	−29.6
Hydrate formation Pressure (barg)	449	1100

Hydrate Formation Control

The following instructions were developed to investigate the simulation results such that no hydrate formation was predicted:

- The minimum outlet temperature from chiller should be −15 °C (the hydrate formation temperature is −18 °C).
- The minimum lean glycol concentration should be 90 weight%.
- The glycol injection rate to heat exchanger (E-402) and propane chiller should be adjusted to 1.1 and 0.3 m³/h respectively.
- The glycol injection pressure should be 10 bar more than the gas pressure.
- Liquid level in filters is controlled at its minimum level.

Optimization of Water and Hydrocarbon Dew Points

In order to optimize dew point temperature and be assured of no hydrate formation, filters and separators were simulated by steady state simulation software (Hysys). The following results were achieved:

- As water was the main reason for hydrate formation, the optimum outlet temperature of gas from heat exchanger (E-401) was determined as 25–30 °C at which the maximum amount of water was separated at the entrance of the unit before the gas temperature reached 0 °C. At this temperature, more than 95% of water was removed and the possibility of hydrate formation decreased considerably.

- Owing to a decrease in the amount of water along with gas stream, the glycol injection rate to the heat exchanger was decreased to 0.3 m³/h and the glycol injection rate to chiller was increased to 1 m³/h.

- Considering these modifications, the gas temperature reduces from 3 °C to −15 °C, however as a consequence of the glycol injection the water dew point reduced from −10 °C to −26 °C as a result of good water separation.

- Owing to a decrease in rich glycol concentration and its lower flow rates, water separation before glycol injection led to an improvement in water dew point (−26 °C) and better warming of liquid hydrocarbons/glycol. Consequently, the glycol separation was performed more effectively which eliminated the glycol losses and fouling problems in the reboilers and heat exchangers of the glycol regeneration unit.

Considering the above mentioned modifications, no freezing problems have been reported which violate product specifications and result in customer dissatisfaction. As shown in Fig. 5, the comparison of Sarkhun gas plant experimental results before and after adjustments represented a considerable improvement in water and hydrocarbon dew points with respect to the cooled gas temperature in propane gas chiller. As seen, the water dew point improved significantly from −10 °C to −26 °C. Also, the hydrocarbon dew point changed from −5 °C to −9 °C and it was almost constant after modifications.

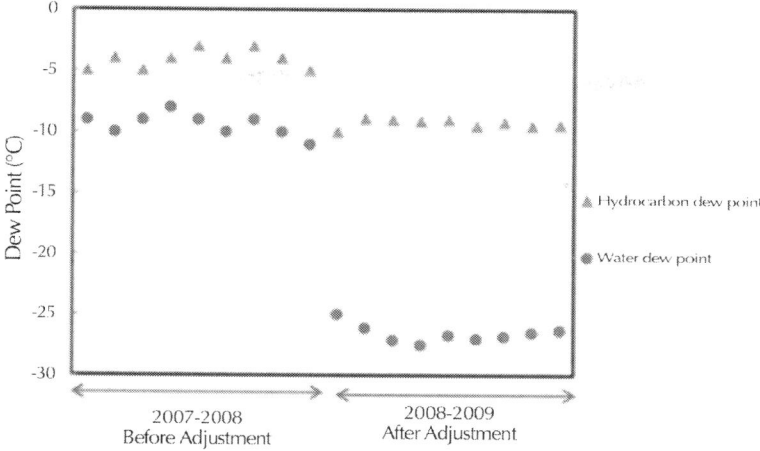

Figure 5: The water and hydrocarbon dew points before and after temperature optimization.

Enhancement LPG Production

The LPG production rate in the Sarkhun gas plant before and after modifications is shown in Fig. 6. Due to the improvement in hydrocarbon dew point and elimination of LPG range hydrocarbons such as C_3 and C_4 from the gas, LPG production rate increased from 145 m³/day to more than 160 m³/day.

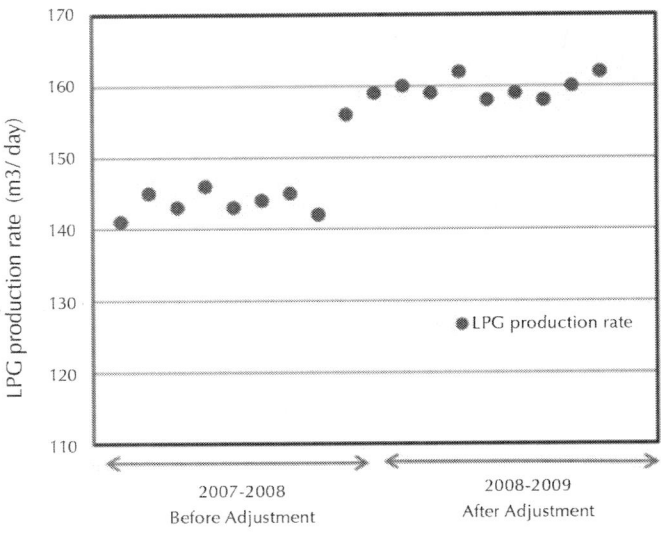

Figure 6: LPG production rate before and after temperature optimization.

Improvement in Glycol Regeneration Unit

Owing to inappropriate adjustment of dew point and low temperature of three phase separator (S-404), liquid hydrocarbons did not separate acceptably from glycol. The presence of liquid hydrocarbons along with glycol caused the following problems in the glycol regeneration unit:

- Increasing pressure drop in lean and rich glycol filters and a failure in their performance which led to changing filters three times a day.

- Increasing pressure drop in glycol exchangers especially the glycol–glycol exchanger. They had to be maintained every three months, if the pressure drop increased to more than 1 bar. The tube cleaning procedure was very difficult.

- Failure in the performance of carbon filters because they were coated with a tar like substance which increased the consumption rate of fresh activated carbon to more than 2500 kg/month.

By eliminating the problems in the dew point adjustment unit and controlling temperature of (S-404) in the range of 25–30 °C, the entrance of liquid hydrocarbon along with glycol to the unit was successfully prevented. Therefore, the consumption rates of filter elements and glycol decreased considerably as shown in Figs. 7 and 8. Furthermore, the problems such as increased pressure drop in exchangers and the presence of tar like substances on the surface of filters, reboiler and shell side of exchangers were completely removed.

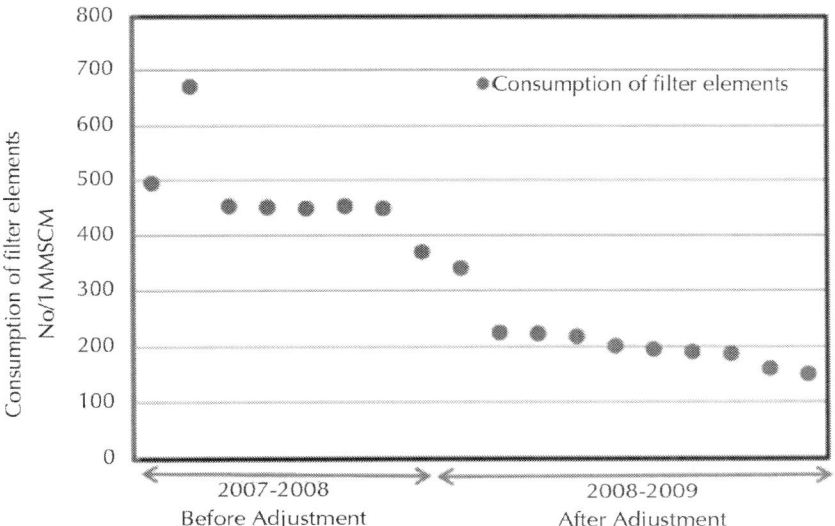

Figure 7: The consumption rates of filter elements before and after temperature optimization.

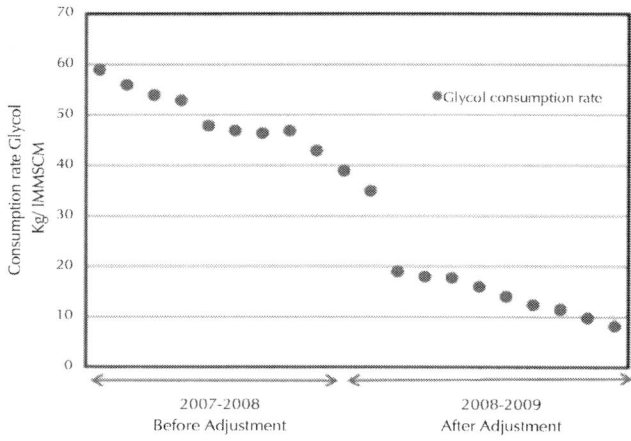

Figure 8: The consumption rates of glycol before and after temperature optimization.

Evaluation of Cost Saving

Table 5 represents the minimum cost saving for the Sarkhun gas plant with more details. As reported in this table, the cost saving by decreasing glycol consumption after modification was about $2.4 million per year. Also decreasing the number of consumed element filters in the dew point adjustment and the glycol regeneration units led to $0.1 million and $0.3 million per year cost saving, respectively. Increasing NGL production was another important achievement which increased total income by $10.3 million per year. It was observed that the annual saving by upgrade of about 4800 MMSCM to sales gas quality is about $484 million by applying all the above mentioned modifications to the Sarkhun gas plant. The net income was calculated without considering the income resulting from prevention of environmental pollutants. Most pollutants were considerably decreased owing to a reduction in gas emission and vaporized LPG venting to atmosphere, filter elements, activated carbon, fouling and tar like substances in glycol regeneration unit. Also, the probable damage to facilities such as analyzer, turbines etc. was not taken into consideration.

Table 5: The minimum cost saving in Sarkhun gas processing

Saving	Before modifica-tion	After modifica-tion	Amount/ number of savings increase production	Unit price based on 2008 years	Cost saving (million$/ year)
Number of Element filter consumption in dew point unit	247	57	190	589$	0.1
Number of Element filter consumption in Glycol regenera-tion unit	4961	2571	2390	117$	0.3
Consumption of glycol (Drum[a])	996 (Drum/ year)	268 (Drum/ year)	728 (Drum/ year)	3267 ($/ Drum)	2.4
Increased NGL Production	–	–	24,000 (ton/year)	430 ($/ton)	10.3
Upgrade of about 4800 MMSCM to sales Gas Quality	–	–	–	–	484
Saving due to maintenance	–	–	–	–	0.01
Total income					497.1

[a](1 Drum) glycol = (230 kg 200 g) glycol.

Environmental Considerations

The following factors caused environmental pollution at the Sarkhun gas plant before modification:

- Repeated changing of filter elements.
- Fouling in exchangers of glycol regeneration unit.
- Gas burning in flare.
- The high consumption rate of activated carbon and no possibility to be regenerated owing to contamination by a tar like substance.

- Glycol losses along with liquid hydrocarbons disposed to burn pits.
- Repeated inspection of separators and exchangers in glycol regeneration and dew point adjustment unit requiring depressurization and purging to flare/atmosphere.
- Unscheduled repairing and vessels inspection.

The above mentioned problems significantly increased greenhouse gas emission to atmosphere. However, these problems were satisfactorily overcome after modifications. One of the outstanding achievements of this research was a remarkable reduction in the greenhouse gas emission to the atmosphere after modifications in Sarkhun gas refinery. The following achievements were obtained after modifications which considerably decreased the environmental pollutions.

Prevention of Greenhouse Gases Production

The downgrading of sales gas product to fuel gas decreased considerably after modifications. Due to inappropriate controlling of liquid level in separators (for instance (S-404)), gas was entrained with liquid hydrocarbons. Therefore, the entrained gases along with liquids and the gas derived from liquid flash which were used as fuel gas in the plant increased considerably. Due to variable composition and consumption rate, the additional produced fuel gas was directed to flare and burned, consequently increasing the production of greenhouse gases. The available information revealed that up to 20% additional gas was burnt in flares in some months. Moreover, the average increase in additional fuel gas was approximated to 15% in each month. It is worth mentioning that some projects have been planned regarding installation of fuel gas compressors in order to compress and inject the additional fuel gas to gas plant sales production with the aim of zero flaring. The decreasing trend of greenhouse gas emission is illustrated in Table 6 and a summarized greenhouse gas emission decline is illustrated in Fig.9.

Table 6: Emissions reduction of greenhouse gas

Year	CO_2(ton/year)	CO (ton/year)	SO_2 (ton/year)	NOX (ton/year)
2007	62,709	54.4	371.9	41.3
2008	3720	39.7	11.8	29.3
2009	36,416	38.8	8.3	28.2

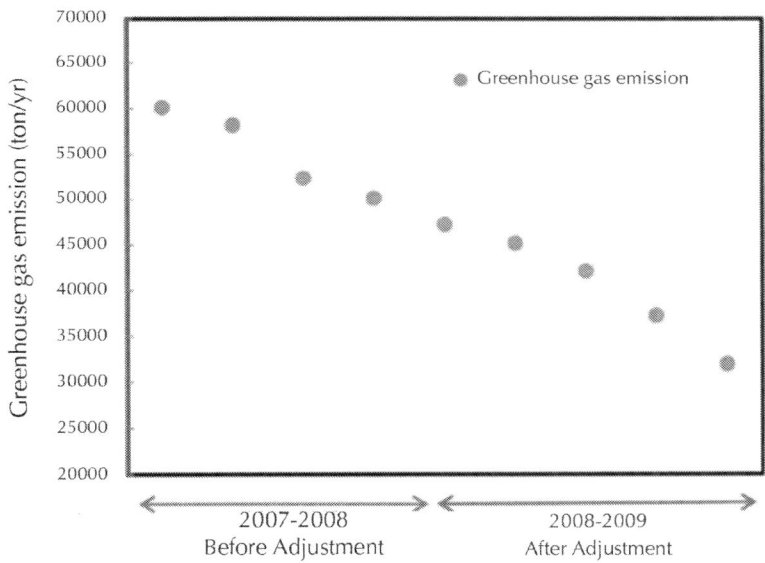

Figure 9: The greenhouse gases emissions before and after modifications.

Prevention of Solid Waste Production

Solid waste appears in different forms such as sediments, waste, slugs and consumed filter elements of dew point adjustment and glycol regeneration units, applied activated carbon of carbon filters of glycol regeneration unit and etc. According to environmental considerations, the solid waste decreased considerably after modifications. As reported in Table 7, solid waste production reduced from 5232 to 1306 kg/year during 2007 until 2009 by applying the above mentioned process modifications.

Table 7: A remarkable decrease in solid waste during the modification period

Year	Solid waste (kg/Year)
2007	5232
2008	1611
2009	1306

Complete Elimination of Natural Gas Emission to Atmosphere

Natural gas (CH_4) emission to atmosphere is very dangerous and harmful. Although prior to repair and opening of the facilities, N_2 purging was used during filters change out in order to prevent the hydrocarbon gas emission to atmosphere, the natural gas emission to atmosphere increased considerably mainly due to process upsets on controlling instruments, inspections and repeated changing of filter elements. However, the probability of hydrocarbon gas emission to atmosphere decreased remarkably and even decreased to zero after modifications, a worthy and great achievement.

CONCLUSIONS

In this study, the natural gas dehydration process of the Sarkhun gas plant was simulated by steady state simulation software and the optimum separation temperature was determined. According to the simulation results, in order to improve processes performance, the glycol concentration should be increased from 85% to 92% and also the glycol flow rate should be increased to 4 m³/h. The experimental results indicated that by these modifications, the water dew point improved well from −10 °C to −26 °C and as a result hydrate formation was prevented. Thus, the LPG production rate increased from 145 m³/day to more than 160 m³/day. Moreover, approximately $500 million annual cost saving was observed, dominated by bringing the product gas into specification for sales. Furthermore, the following modifications were recognized:

- Increasing the injection pressure from 75 bar to 80 bar.
- Increasing the glycol concentration from 82 to 92 weight%.
- Modifying the glycol injection location from gas–gas heat exchanger to propane chiller.

By applying above modifications in Sarkhun gas refinery, the water dew point temperature decreased well to −26 °C. One of the outstanding achievements of this research was a considerable decrease in greenhouse gas emission to the atmosphere, mainly due to a remarkable decrease in exchanging of filter elements, consumption rate of activated carbon, gas burning in flare and fouling in exchangers of glycol regeneration unit.

REFERENCES

1. Acor, L.G., Mirdadian, D., 2003. Benefits of using deliquescinq desiccants for gas dehydration. In: Proceedings e SPE Production Operations Symposium, pp. 669e674.

2. Ale, R.L., Mercado, L., Tarifa, E., Erdmann, E., 2009. Natural gas dew point adjustment: parametric sensitivity analysis. In: 8th World Congress of Chemical Engineering: Incorporating the 59th Canadian Chemical Engineering Conference and the 24th Interamerican Congress of Chemical Engineering.

3. Branco, D.A.C., Szklo, A.S., Schaeffer, R., 2010. CO2e emissions abatement costs of reducing natural gas flaring in Brazil by investing in offshore GTL plants producing premium diesel. Energy 35, 158e167.

4. Campbell, J.M., 1992. Gas Conditioning and Processing, seventh ed.. In: Campbell Petroleum Series Oklahoma.

5. Choi, M.S., Spisak, C.D., 1993. Aromatic recovery unit (ARU): a process enhancement for glycol dehydrators. In: Proceedings of the SPE/EPA Exploration and Production Environmental Conference, pp. 199e205.

6. Covington, K., Lyddon, L., Ebeling, H., 1998. Reduce emissions and operating costs with appropriate glycol selection. In: Proceedings, Annual Convention e Gas Processors Association, pp. 42e48.

7. Darwish, N.A., Hilal, N., 2008. Sensitivity analysis and faults diagnosis using artificial neural networks in natural gas TEG dehydration plants. Chem. Eng. J. 137, 189e 197.

8. Faulkner, L.L., 2006. Fundamentals of Natural Gas Processing. Taylor and Francis Group, LLC, New York.

9. Grizzle, P.L., 1993. Hydrocarbon emission estimates and controls for natural gas glycol dehydration units. In: Proceedings of the SPE/EPA Exploration and Production Environmental Conference, pp. 177e186.

10. Hicks, R., Gallaher, D., Craig, R., 2004. Processing: pressurized reboiler reduces VOC emissions in glycol dehy systems. Oil Gas J. 102, 90e96.

11. Katz, D.L.V., 1959. Hand Book of Natural Gas Engineering, first ed. McGraw-Hill Book Company.

12. Khosravanipour, A.M., Rahimpour, M.R., Shariati, A., 2009. Vapor-liquid Equilibria of WaterþTriethylene glycol (TEG) and WaterþTEGþToluene at 85 kPa. J. Chem. Eng. Data 54, 876e881.

13. Kohl, L., Nielsen, R.B., 1997. Gas Purification, fifth ed. Gulf Publishing Company, Houston.

14. Løkken, T.V., Bersås, A., Christensen, K.O., Nygaard, C.F., Solbraa, E., 2008. Water content of high pressure natural gas: data, prediction and experience from field. In: International Gas Research Conference Proceedings, vol. 3, pp. 1979e2021.

15. Mokhatab, S., Poe, W.A., 2006. Handbook of Natural Gas Transmission and Processing. Gulf Professional Publishing.

16. Paymooni, K., Rahimpour, M.R., Raeissi, S., Abbasi, M., Saviz, M.B., 2011. Enhancement in triethylene glycol (TEG) purity via hydrocarbon solvent injection to a TEG þ water system in a batch distillation column. Energ. Fuel 25, 5126e5137.

17. Pearce, R.L., Protz, J.E., Lyon, G.W., 1972. Dry gas to low dew points. Hydrocarbon Process 51, 79e81.

18. Peinemann, K.V., Stegger, J., 2003. New membranes and processes for the separation of gases and vapors. Chem.-Ing-Tech 75, 1159e1160.

19. Rahimpour, M.R., Seifi, M., Paymooni, K., Shariati, A., Raeissi, S., 2011. Enhancement in NGL production and improvement in water dew point temperature by optimization of slug catchers' pressures in water dew point adjustment unit. J. Nat. Gas Sci. Eng. 3, 326e333.

20. Rahimpour, M.R., Jokar, S.M., Feyzi, P., Asghari, R., 2013. Investigating the performance of dehydration unit with coldfinger technology in gas processing plant. J. Nat. Gas Sci. Eng. 12, 1e12.

21. Rojey, A., Jaffret, C., 1997. Gas: Production Processing, Transport, first ed. Inst. Fracais Du Petrole Pub, Paris.

22. Rueter, C.O., Ogle, L.D., Reif, D.L., Evans, J.M., 1993. Development of sampling and analytical methods for measuring BTEX and VOC from glycol dehydration units. In: Proceedings of the SPE/EPA Exploration and Production Environmental Conference, pp. 125e134.

23. Twu, C.H., Tassone, V., Sim, W.D., Watanasiri, S., 2005. Advanced equation of state method for modeling TEGs water for glycol gas dehydration. Fluid Phase. Equilib. 228e229, 213e221.

Chapter 3

Titanium Dioxide Nanofibers and Microparticles Containing Nickel Nanoparticles

Faheem A. Sheikh[1], Javier Macossay[1], Muzafar A. Kanjwal[2], Abdalla Abdal-hay[3, 4], Mudasir A. Tantry[5], and Hern Kim[6]

[1]Department of Chemistry, University of Texas-Pan American, Edinburg, TX 78539, USA

[2]DTU Food, Technical University of Denmark, Soltofts Plads, Building 227 2800 Kgs. Lyngby, Denmark

[3]Department of Bio and Nano System Engineering, College of Engineering, Chonbuk National University, Jeonju 561-756, Republic of Korea

[4]Department of Mechanical Design and Materials Engineering, Chonbuk National University, Jeonju 561-756, Republic of Korea

[5]National Center for Natural Products Research, Research Institute of Pharmaceutical Sciences, School of Pharmacy, the University of Mississippi, Oxford, MS 38677, USA

[6]Department of Environmental Engineering and Biotechnology, Energy & Environment Fusion Technology Center, Myongji University, Kyonggi-do, Yongin 449-728, Republic of Korea

ABSTRACT

The present study reports on the introduction of various nanocatalysts containing nickel (Ni) nanoparticles (NPs) embedded within TiO_2 nanofibers and TiO_2 microparticles. Typically, a sol-gel consisting of titanium isopropoxide and Ni NPs was prepared to produce TiO_2 nanofibers by the electrospinning process. Similarly, TiO_2 microparticles containing Ni were prepared using a sol-gel syntheses process. The resultant structures were studied by SEM analyses, which confirmed well-obtained nanofibers and microparticles. Further, the XRD results demonstrated the crystalline feature of both TiO_2 and Ni in the obtained composites. Internal morphology of prepared nanofibers and microparticles containing Ni NPs was characterized by TEM, which demonstrated characteristic structures with good dispersion of Ni NPs. In addition, the prepared structures were studied as a model for hydrogen production applications. The catalytic activity of the prepared materials was studied by in situ hydrolysis of $NaBH_4$, which indicated that the nanofibers containing Ni NPs can lead to produce higher amounts of hydrogen when compared to other microparticles, also reported in this paper. Overall, these results confirm the potential use of these materials in hydrogen production systems.

INTRODUCTION

In recent years, due to concerns about global warming and the depletion of fossil fuels from the natural reservoirs, the utilization of various other sources of energy had been intensively investigated by scientific society. There are various means of obtaining energy from the natural and artificial resources. Among the various forms of energy, hydrogen has become one of the most promising future energy means of harvesting. However, the production of this important source by direct water splitting without any byproducts is one of potential alternatives to hydrogen fuel for future energy supply [1, 2]. In order to overcome this rising demand for hydrogen, many methods have been devised, such as reforming of natural gas [3, 4], coal gasification [5], biomass pyrolysis and gasification [6], hydrolysis of chemical hydrides [7, 8], and electrolytic or photocatalytic water splitting [1, 2]. Although the hydrogen production by water splitting using electrolysis of alkaline solution is commercially done, the efficiency of the process is low. However, efficiencies are increased through the use of polymer electrolyte membranes and photovoltaic reactions. Recently, there has been a growing interest in hydrogen generation and storage using metal hydrides, such as lithium hydride (LiH) [9], sodium aluminum hydride ($NaAlH_4$) [10], lithium borohydride ($LiBH_4$) [11], and sodium borohydride ($NaBH_4$) [12–14]. $NaBH_4$ is the most favorable compound for hydrogen production because of its high hydrogen density, stability in alkaline solution [15], pure hydrogen generation [16], and recycling of the byproducts [17]. The governing reaction for hydrogen storage and generation is given as

$$NaBH_4 + 2H_2O \longrightarrow NaBO_2 + 2H_2 \Delta H = 300\,KJ/mol. \tag{1}$$

Similarly, the strategies to use the metal-catalyst in presence of $NaBH_4$ for accelerating the rate of hydrogen production had recently been accomplished. These various metal-catalysts include the platinum (Pt) [18], palladium (Pd) [19], ruthenium (Ru) [20], cobalt (Co) [21], Co-B [22], nickel (Ni) [23], Ni-B [13], Ni-Co-B [24], and carbon nanotubes (CNT) [25] had been extensively utilized for the hydrogen production. Nickel metal is considered economically promising compared with other catalyst-forms especially with that of platinum. This is the reason the use of nickel-based catalysts had been evaluated for the production of hydrogen [23, 24]. It is noteworthy to mention that Ni metal in the form of nanoparticulate in pure or associated form can be effactually used to boost the production of hydrogen when we used $NaBH_4$ [23, 24]. Having these essential properties, we have selected nickel nanoparticles and $NaBH_4$ as model catalyst for hydrogen production for the present study.

The electrospinning technique has attracted considerable attention due to the production of fibers with diameters that range from the micrometer to the nanometer size [26, 27]. In a typical electrospinning process, an electrostatically driven polymer jet is ejected from a polymer solution or a sol-gel which undergoes a bending instability wherein the solvent evaporates, and an ultrafine stretched fiber is deposited on a grounded collector [28]. Consequently, the nanofibers obtained by this technique possess large surface areas when compared to other nanoparticle forms [29].

The present work presents the fabrication of various kinds of nanocatalysts forms and their capability to produce hydrogen. The fabricated nanoforms were fabricated as pure TiO_2 nanofibers, modified TiO_2 nanofibers containing Ni nanoparticles (NPs) and TiO_2 microparticles containing Ni NPs. These prepared nanocatalysts have been intensively studied and well characterized with various states of the art techniques. After characterization, the efficiency of these materials was tested for the production of hydrogen through in situ hydrolysis of $NaBH_4$.

EXPERIMENTAL SECTION

Materials

Poly (vinyl acetate) (PVAc, Mw = 500,000? g/mol) was obtained from Sigma Aldrich, USA. Titanium (IV) isopropoxide [Ti (Iso)], 98% assay was purchased from Junsei Co. Ltd., Japan. Nickel nanopowder <100? nm, 99.9% pure was purchased from Aldrich, USA. N, N-dimethylformamide (DMF) was obtained from Showa Chemicals Ltd., Japan and used without further purification.

Characterization

The morphology of the obtained nanocatalysts was analyzed utilizing a JEOL JSM-5900 scanning electron microscope, JEOL Ltd., Japan. The phase and crystallinity of the nanofibers and microparticles was investigated using an X-ray diffractometer (XRD, Rigaku Co., Japan) with Cu Ka (= 1.540 Å) radiation over a Bragg angle ranging from 20 to 80°. Transmission electron microscopy (TEM) was done with a JEOL JEM 2010 operating at 200? kV, JEOL Ltd., Japan.

Procedure

Fabrication of Nanofibers by Electrospinning

The electrospinning process was utilized to produce TiO_2 nanofibers containing Ni NPs. Typically, a sol-gel was prepared by mixing Ti (Iso) and PVAc (20? wt%, in DMF) with a weight ratio of 2? 3. Thereafter, a few drops of acetic acid were added until the solution became transparent under stirring. To fabricate the sol-gels containing Ni NPs, a step by step methodology was adopted. Briefly, 0.5? g of nickel NPs, were added into a previously prepared transparent solution of Ti(Iso)/PVAc; the solution was subsequently homogenized under stirring for 10 minutes. A high voltage power supply (CPS-60 K02V1, Chungpa EMT Co., Republic of Korea), capable of generating voltages up to 60? kV, was used for electrospinning the sol-gels. The solution to be electrospun was supplied through a plastic syringe attached to a capillary tip, which contained a copper pin to connect to the positive electrode (anode) in the high power supply. The electrospinning system was completed through the attachment of the negative electrode (cathode) to a grounded metallic collector. The nanofibers were deposited on rectangular collector covered with thin sheet of aluminum foil, equipped with heating system having temperature of 40°C, which helps to remove the residual solvents after the fiber lands on collector (Scheme 1). The solutions were electrospun at 15? kV and 15? cm working distance (the distance between the needle tip and the collector). The as-spun fibers were initially dried for 24? h at 80°C under vacuum in the presence of P_2O_5, to remove the remaining residual solvents. In order to remove the polymer used in making sol-gels, the samples were additionally heated in air atmosphere at 600°C for 1?h with heating rate of 5°C/min.

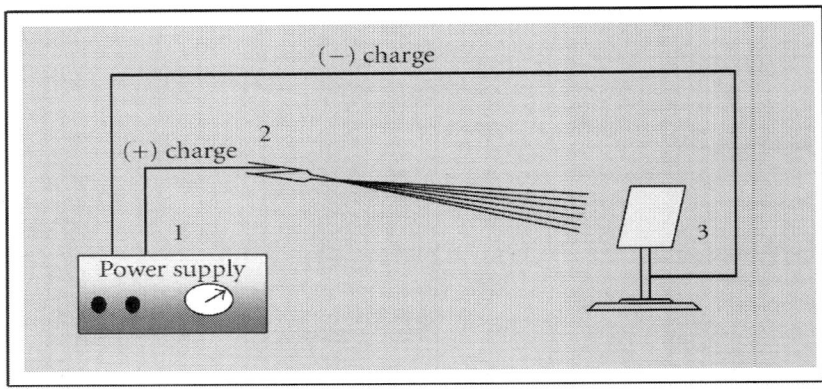

Scheme 1: The schematic illustration of a simple electrospinning spinning apparatus: (1) dc power supply (2) syringe, and (3) collector.

Fabrication of Microparticles by Sintering

A sol-gel was prepared by mixing Ti (Iso) and PVAc (20 wt% in DMF) with a weight ratio of 2:3, respectively. Thereafter, a few drops of concentrated acetic acid were added under stirring to afford a transparent solution, to which 0.5g of nickel NPs were added. Hereafter, these solutions were homogeneously mixed under stirring for 10 minutes. However, instead of electrospinning as previously described for fabrication of nanofibers, this solution was dried under vacuum at 80°C for 48 h to completely remove the solvents. The obtained solid materials were finely ground and sintered in air at 600°C for 1 h with a heating rate of 5°C/min. After the sintering process, the samples were further subjected to fine grinding to additionally reduce their size.

Hydrogen Production Studies

All the samples were investigated for their catalytic activity for the in situ hydrolysis of $NaBH_4$. Typically, 50 mg of all the sample combinations were placed in a specially designed tight sealed flask which contained 50 mL of a distill water containing 50 mg of $NaBH_4$ at constant temperature (25°C). The catalytic performances were compared for the hydrogen production from hydrolysis of $NaBH_4$. Briefly, the reaction proceeded at a stirring rate of (1000?rpm) and the amount of hydrogen generated over time was measured immediately after all the components were added through the water displacement method [30], where the volume of hydrogen is equal to that of the displaced water whose weight was recorded by a balance. The balance connected with a computer, which was installed with the software of "Balance Talk", can read the weight of balance automatically.

RESULTS AND DISCUSSIONS

In these experiments, PVAc was used as a binder for making the sol-gels to get good viscous solutions, so as to have the appropriate bending instability during the electrospinning process. After getting the nanofiber mats from electrospinning of the sol-gels, the obtained nanofiber mats were dried, and further subjected to frying in a furnace to remove the polymer binder (i.e., PVAc). In this context, (figure 1(a)) shows the SEM images of nanofibers obtained after performing the frying process. As shown in this figure, the morphology of pristine nanofibers (after calcination) consists of pure TiO_2 and is not affected by the high temperature heating. According to the high temperature of calcination used at 600°C, which is twice the thermal degradation temperature of PVAc [31], this temperature should have been sufficient to remove the PVAc completely, leaving behind only ceramic TiO_2. As can be observed in this image, that nanofiber morphology is in accordance with our previously established reports [32]. Furthermore, this image allows us to determine the average diameters of obtained nanostructures by using (Photoshop 6), around (10–12 individual diameters were measured per sample). The resultant diameters in case of pure TiO_2 nanofiber were to be 600 ±320? nm.

(a)

(b)

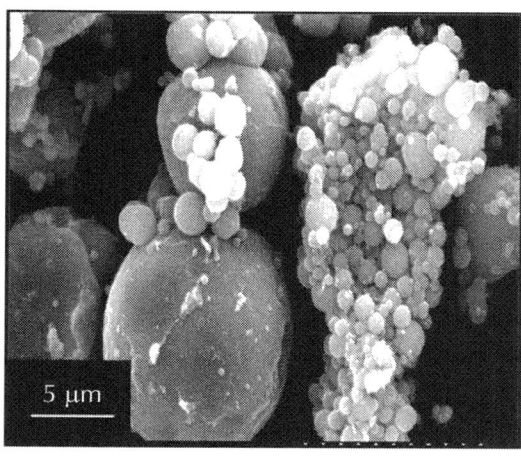

(c)

Figure 1: SEM Images of the obtained catalysts after the calcinations process. Images of the pristine TiO$_2$ nanofibers (a), images of the TiO$_2$ nanofibers containing Ni NPs (b), and TiO$_2$ microparticles containing Ni NPs (c).

Figure 1(b) shows the SEM morphology of the nanofibers containing Ni NPs. It is observed that the morphology of the nanofibers was not affected by the addition of the Ni NPs, if the nanofibrous morphology is considered. However, an important observation was observed that nanofibers containing Ni presented a smaller diameter than the pristine TiO$_2$ nanofibers. Moreover, the average fiber diameter calculated was to be 982±391? nm for nanofibers containing Ni NPs, which is comparatively less than (600±320? nm) as that was seen in case of pristine TiO$_2$ nanofibers. It is likely that the colloidal solutions used for electrospinning were highly conductive after addition of Ni, thus promoting a reduction in the fiber diameter [33]. A schematic representation for the fabrication of nanofibers is produced in (Scheme 2). Briefly, the first step of this scheme includes the preparation of a sol-gel containing Ni NPs. After this step, the prepared sol-gel is electrospun and calcinated to remove the binder (PVAc). The final step results in the formation of TiO$_2$ nanofibers containing Ni NPs.

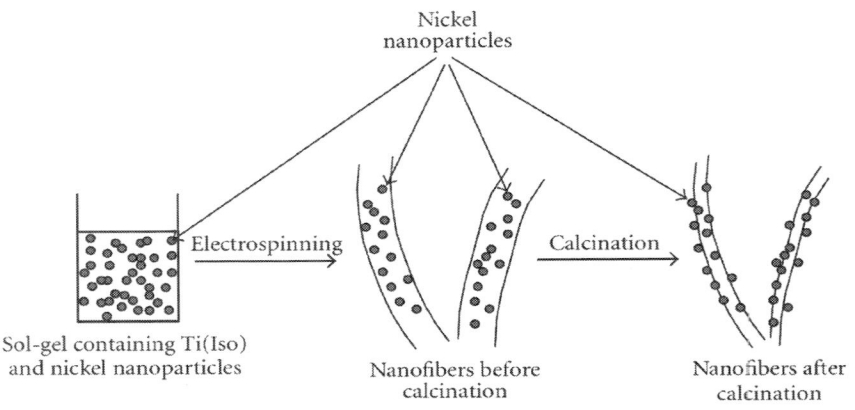

Scheme 2: Representation of this novel strategy to fabricate TiO_2 nanofibers containing Ni NPs.

SEM results of the TiO_2 microparticles containing Ni NPs are shown in (figure 1(c)). It is observed that spherical particles are predominately present with wide range of diameters. The occurrence of particles appears in the form of clusters with the average diameters of (7000±3607? nm). To verify the addition of Ni nanoparticles into the nanofibers and microparticles, SEM-EDX analysis was performed on the obtained structures (Figure 2). As shown in (Figure 2(a)) and its corresponding EDX data from the area analyses of the nanofiber, the presence of Ti from this image clearly corresponds to the occurrence of pristine TiO_2 nanofibers. The EDX data of its counterpart which contains certain amount of Ni in the nanofibers is also present in (Figure 2(b)). From this figure we can clearly find, that in addition to the peaks of Ti, there is obvious presence of Ni peaks. This spectrum clearly demonstrates that small-sized NPs can be embedded in the nanofibres. Besides, this it can be clearly seen that diameter of these nanofibers containing Ni NPs is comparatively less than the pristine nanofibers (i.e., Figure 2(a)). This observation about the SEM-EDX results corroborates the simple SEM (Figures 1(a) and 2(b)) results, which elucidate the decrease in size of nanofibers after the addition of conductive Ni NPs. The EDX data originating from the TiO_2 microparticles containing Ni NPs is presented in (Figure 2(c)). In this figure, the peaks corresponding to the TiO$_2$ and Ni are present.

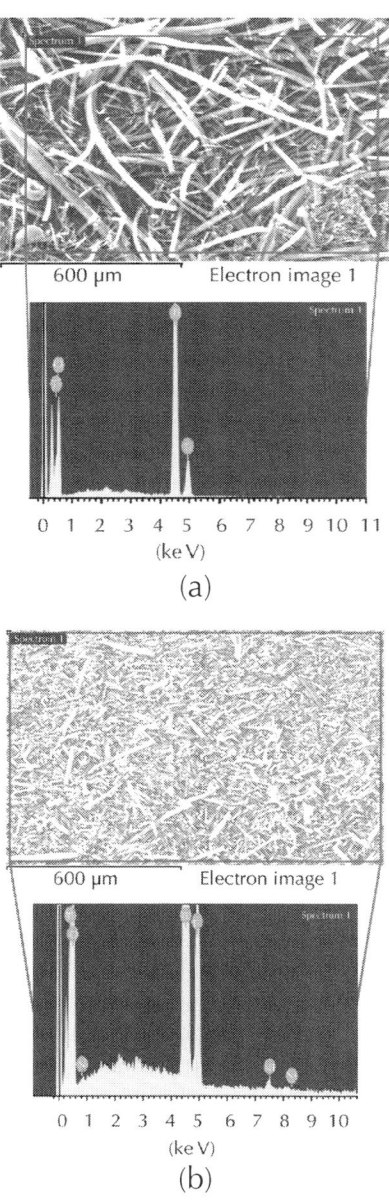

(a)

(b)

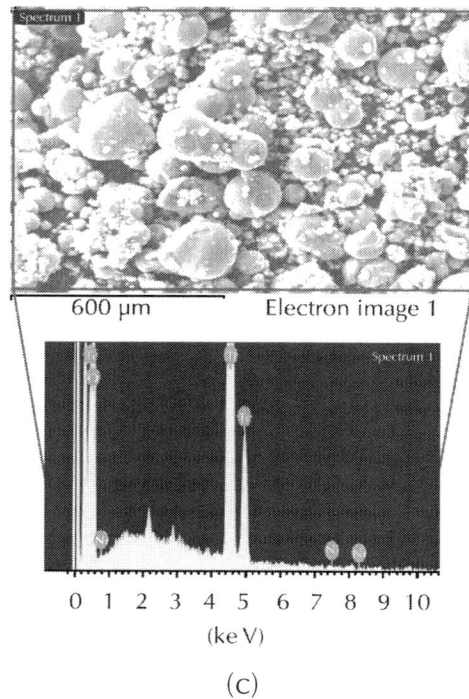

600 μm   Electron image 1

0 1 2 3 4 5 6 7 8 9 10
(ke V)

(c)

Figure 2: SEM images with EDX analysis. The EDX area of the pristine TiO$_2$ nanofibers and its corresponding data (a), the EDX area of the modified TiO$_2$ nanofibers containing Ni NPs and its corresponding data (b), the EDX area of the modified TiO$_2$ microparticles containing Ni NPs and its corresponding data (c).

It is well known, that TEM can be utilized to differentiate between two different materials in regards to their different crystalline patterns. Therefore, to investigate the crystalline features of the prepared materials, TEM images were obtained and presented in (Figure 3). Figure 3(a) shows the results from pure TiO$_2$ nanofibers, where it can be seen that the morphology of individual nanofibers is consistent with that of the defect-free morphology obtained from SEM images (Figure 1(a)). The high resolution transmission electron microscope (HR-TEM) image of the pristine nanofibers is presented in (Figure 3(b)). The portion indicated by arrow in the HR-TEM image indicates that there is no dislocation of the crystal lattice and the crystal planes are arranged in a linearly unique pattern consistent of pure TiO_2 [34]. Figure 3(c) presents TEM images of the nanofibers containing Ni NPs at low magnification, where it is confirmed that the nanoparticles are present on the nanofiber surfaces. These Ni NPs can be seen as darker areas than the main TiO_2 nanofiber, as indicated by the arrows. HRTEM images of the marked areas in (Figure 3(c)) are presented in (Figure 3(d)). Overall, these images reveal the expected cubic shape for the Ni NPs, with diameters from 15 to 20nm, is present. (Figure3 (e)) represents the low magnification TEM image of the microparticles containing Ni NPs. The arrow originating from encircled area shows the HR-TEM of the Ni NPs in (Figure 3(f)). It can be clearly seen, that Ni NPs are well associated with TiO_2 microparticles. In these figures, it can be seen that crystals have good atomic arrangement with respect to two components. The atoms can be seen as uniformly arranged having regular and periodic behavior indicating individual TiO_2 and Ni components in them, which overall indicate good crystallinity of the synthesized structures. Therefore, one can come up to an interesting finding that the crystal patterns of the both components are arranged in a leaner format and are without the dislocations or imperfections in the lattice planes. It is noteworthy to mention that in case of SEM analyses the presence of Ni NPs in case of nanofibers and microparticles was not so obvious, rationale is due high intensity electron beam of TEM which can differentiate the internal as well as external contents of the crystalline materials, of which former one is incapable to do alone, due to its poor resolution.

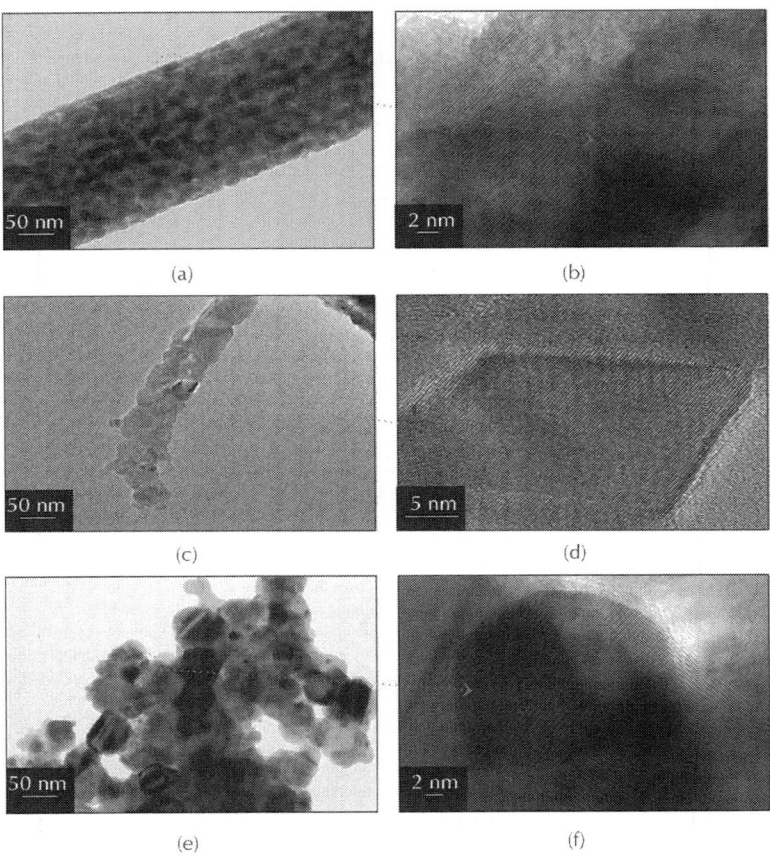

(a)

(b)

(c)

(d)

(e)

(f)

Figure 3: TEM images of nanofibers and microparticles after calcination process. Pristine TiO$_2$ nanofibers in low magnification (a), the HRTEM images of the former figure (b). The low magnification images of the TiO$_2$ nanofibers containing Ni NPs (c), the HR-TEM images of the corresponding former figure. The low magnification images of the TiO$_2$ microparticles containing Ni NPs (e), the HR-TEM images of the corresponding figure (f).

As shown in (Figure 4), the spectra show the XRD pattern of pristine TiO$_2$ nanofibers, TiO$_2$ nanofibers containing Ni NPs, and TiO$_2$ microparticles containing Ni NPs. In all of the materials obtained, there are strong diffraction peaks at 2? values of 25.50, 37.69, 48.04, 53.85, 55.08, 62.75, and 79.98°, which correspond to the crystal planes (110), (401), (020), (601), (513), (403), and

(621), respectively, and indicate the formation of pure anatase titanium dioxide [35]. Further, the TiO$_2$ nanofibers containing Ni NPs and TiO$_2$ microparticles containing Ni NPs presented the TiO$_2$ peaks as well as the additional peaks attributed to Ni at 2? values of 27.50, 36.80, 38.35, 43.16, 62.75, 70.33, and 79.98° corresponding to the crystal planes (220), (311), (311), (312), and (322) [36], thus confirming the presence of the embedded Ni NPs. It is interesting to note that the intensities of the Ni peaks in these composites were the same, which to some extent is expected, since the metal concentrations used were the same.

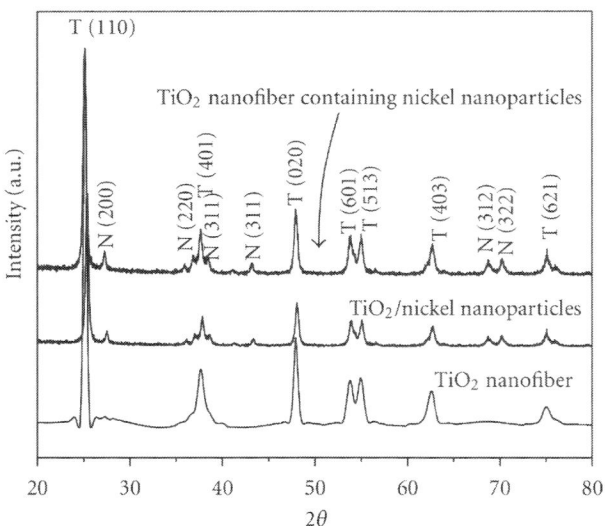

Figure 4: XRD results of the different nanocomposites.

Figure 5 shows the results obtained after using these synthesized structures as nanocatalysts for hydrogen production. These results demonstrate the capability of all prepared catalyst combination render to produce different amounts of hydrogen. Overall, the hydrogen production was in the range of 25 to 250 grams depending upon the nanocatalyst investigated. In more details, it is observed that the highest amount of hydrogen production is shown by the TiO$_2$ nanofibers containing Ni NPs, followed by TiO$_2$ microparticles containing Ni NPs, and finally by the pristine nanofibers. The main reason for the higher amount of hydrogen production from the first two nanocatalysts is clearly attributed due to the presence of Ni NPs [13]. The amount of hydrogen produced by the pristine TiO_2 nanofibers (Ni free) is due to the presence of the original $NaHB_4$ in the processing media (without catalyst). Moreover, the TiO_2 nanofibers containing Ni NPs produce more hydrogen than its TiO_2 microparticle containing Ni NPs analog. It is proposed that the higher surface to volume ratio present in the nanofibers is responsible for this effect [29].

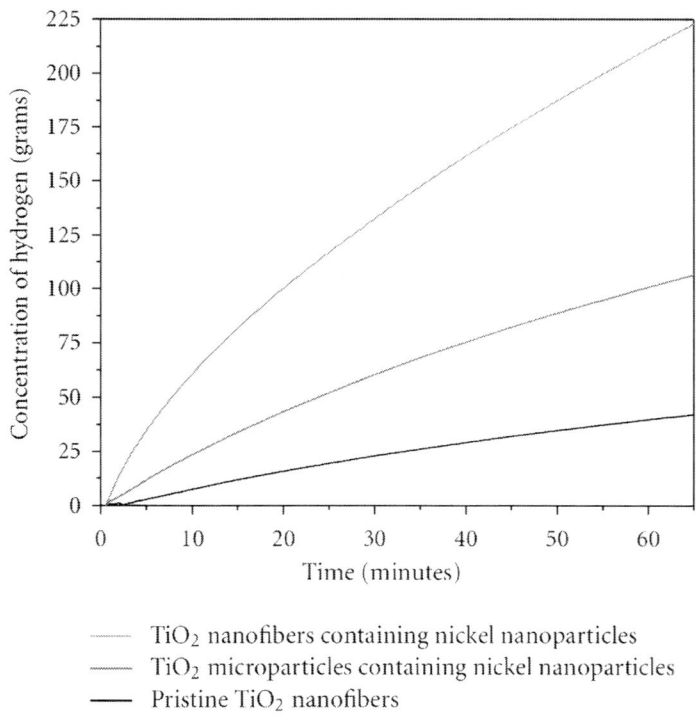

-------- TiO_2 nanofibers containing nickel nanoparticles
———— TiO_2 microparticles containing nickel nanoparticles
———— Pristine TiO_2 nanofibers

Figure 5: Hydrogen production of the by various nanocomposites at 26°C in pH 7.4 distilled water to demonstrate the highest hydrogen production.

To practically investigate that nanofibers do have higher surface to volume ratio than that of microparticles, which in turn can put more light, the surface area of both fibers and microparticles has been measured by using Brunauer-Emmett-Teller (BET) technique (ASAP 2010, Micromeritics, Norcross, GA). It is noteworthy to mention that exact amount of 0.5g of Ni NP was used to mix with sol-gel to fabricate nanofibers and microparticles, therefore one would assume that two combinations would produce same results. However, from those tests, it was observed that microparticles containing Ni NPs had surface area of $13.6765 \pm 0.675 m^2/g$. The pure TiO_2 nanofibers had a surface area of $15.3456 \pm 0.1751 m^2/g$, and the nanofibers containing Ni NPs had a surface area of $23.2782 \pm 0.1961? m^2/g$, respectively. These results are in accordance with our previous reports [29]. Overall, these numbers indicate that nanofibers with Ni had much higher surface area, followed by pure TiO_2 nanofibers and TiO_2 microparticles containing Ni NPs. Furthermore, this higher surface to volume ratio in case of TiO_2 nanofibers containing Ni NPs self explains the reason for more hydrogen production than the other two combinations.

CONCLUSIONS

In conclusion, we were able to fabricate three different types of nanomaterials through combination of sol-gel and electrospinning process. Electrospinning of a colloid comprised of Ti (Iso) and Ni NPs produced ceramic nanofibers that contained attached Ni NPs and partially captured NPs. The SEM instrument was used to find out the morphologies of nanomaterials after the electrospinning and calcinations of sol-gel. SEM equipped with EDX technique was used to differentiate between pristine and Ni-loaded nanofibers and microparticles composites. TEM images were used to determine the appearance of Ni NPs over the individual nanofibers. All the materials obtained were evaluated for their catalytic activity towards hydrogen production, resulting in the nanofibers containing Ni NPs being the most promising materials for this application was due to high surface to volume ratio of nanofibers containing Ni NPs resulted to produce the highest production of hydrogen.

ACKNOWLEDGMENTS

This work was supported by Priority Research Centers program through the National Research Foundation of Korea (NRF) funded by the Ministry of Education, Science and Technology (2012-0006693). F. A. Sheikh and J. Macossay are thankful for partial financial support for this work from NIH-NIGMS-NIA Grant no. 1SC2AG036825-01.

REFERENCES

1.   H. de Battista, R. J. Mantz, and F. Garelli, "Power conditioning for a wind-hydrogen energy system," Journal of Power Sources, vol. 155, no. 2, pp. 478–486, 2006.

2.   H. Miland, R. Glöckner, P. Taylor, R. Jarle Aaberg, and G. Hagen, "Load control of a wind-hydrogen stand-alone power system," International Journal of Hydrogen Energy, vol. 31, no. 9, pp. 1215–1235, 2006.

3.   R. J. Farrauto, "Introduction to solid polymer membrane fuel cells and reforming natural gas for production of hydrogen," applied Catalysis B, vol. 56, no. 1-2, pp. 3–7, 2005.

4.   Heinzel, B. Vogel, and P. Hübner, "Reforming of natural gas—hydrogen generation for small scale stationary fuel cell systems," Journal of Power Sources, vol. 105, no. 2, pp. 202–207, 2002.

5.   G. J. Stiegel and M. Ramezan, "Hydrogen from coal gasification: an economical pathway to a sustainable energy future," International Journal of Coal Geology, vol. 65, no. 3-4, pp. 173–190, 2006.

6.   E. E. Iojoiu, M. E. Domine, T. Davidian, N. Guilhaume, and C. Mirodatos, "Hydrogen production by sequential cracking of biomass-derived pyrolysis oil over noble metal catalysts supported on ceria-zirconia," Applied Catalysis A, vol. 323, pp. 147–161, 2007. · ·

7. S. C. Amendola, S. L. Sharp-Goldman, M. S. Janjua et al., "Safe, portable, hydrogen gas generator using aqueous borohydride solution and Ru catalyst," International Journal of Hydrogen Energy, vol. 25, no. 10, pp. 969–975, 2000.

8. J. H. Kim, K. T. Kim, Y. M. Kang et al., "study on degradation of filamentary Ni catalyst on hydrolysis of sodium borohydride," Journal of Alloys and Compounds, vol. 379, no. 1-2, pp. 222–227, 2004.

9. J. J. Vajo, S. L. Skeith, F. Mertens, and S. W. Jorgensen, "Hydrogen-generating solid-state hydride/hydroxide reactions," Journal of Alloys and Compounds, vol. 390, no. 1-2, pp. 55–61, 2005.

10. J. O. Jensen, Q. Li, R. He, C. Pan, and N. J. Bjerrum, "100–200 °C polymer fuel cells for use with $NaAlH_4$," Journal of Alloys and Compounds, vol. 404–406, pp. 653–656, 2005.

11. Y. Kojima, K. I. Suzuki, and Y. Kawai, "Hydrogen generation from lithium borohydride solution over nano-sized platinum dispersed on $LiCoO_2$," Journal of Power Sources, vol. 155, no. 2, pp. 325–328, 2006.

12. H. I. Schlesinger, H. C. Brown, A. E. Finholt, J. R. Gilbreath, H. R. Hoekstra, and E. K. Hyde, "Sodium borohydride, its hydrolysis and its use as a reducing agent and in the generation of hydrogen,"Journal of the American Chemical Society, vol. 75, no. 1, pp. 215–219, 1953. ·

13. D. Hua, Y. Hanxi, A. Xinping, and C. Chuansin, "Hydrogen production from catalytic hydrolysis of sodium borohydride solution using nickel boride catalyst," International Journal of Hydrogen Energy, vol. 28, no. 10, pp. 1095–1100, 2003.

14. M. Zahmakiran and S. Özkar, "Water dispersible acetate stabilized ruthenium(0) nanoclusters as catalyst for hydrogen generation from the hydrolysis of sodium borohyride," Journal of Molecular Catalysis A, vol. 258, no. 1-2, pp. 95–103, 2006.

15. J. A. Ritter, A. D. Ebner, J. Wang, and R. Zidan, "Implementing a hydrogen economy," Materials today, vol. 6, no. 9, pp. 18–23, 2003.

16. J. H. Kim, H. Lee, S. C. Han, H. S. Kim, M. S. song, and J. Y. Lee, "Production of hydrogen from sodium borohydride in alkaline solution: development of catalyst with high performance,"International Journal of Hydrogen Energy, vol. 29, no. 3, pp. 263–267, 2004.

17. D. L. Calabretta and B. R. Davis, "Investigation of the anhydrous molten Na–B–O–H system and the concept: electrolytic hydriding of sodium boron oxide species," Journal of Power Sources, vol. 164, no. 2, pp. 782–791, 2007.

18. Y. Kojima, K. I. Suzuki, K. Fukumoto et al., "Development of 10 kW-scale hydrogen generator using chemical hydride," Journal of Power Sources, vol. 125, no. 1, pp. 22–26, 2004.

19. N. Patel, B. Patton, C. Zanchetta et al., "Pd–C powder and thin film catalysts for hydrogen production by hydrolysis of sodium borohydride," International Journal of Hydrogen Energy, vol. 33, no. 1, pp. 287–292, 2008.

20. J. S. Zhang, W. N. Delgass, T. S. Fisher, and J. P. Gore, "Kinetics of Ru-catalyzed sodium borohydride hydrolysis," Journal of Power Sources, vol. 164, no. 2, pp. 772–781, 2007.

21. W. Ye, H. Zhang, D. Xu, L. Ma, and B. Yi, "Hydrogen generation utilizing alkaline sodium borohydride solution and supported cobalt catalyst," Journal of Power Sources, vol. 164, no. 2, pp. 544–548, 2007.

22. S. U. Jeong, R. K. Kim, E. A. Cho et al., "A study on hydrogen generation from $NaBH_4$ solution using the high-performance Co–B catalyst," Journal of Power Sources, vol. 144, no. 1, pp. 129–134, 2005.

23. O. Metin and S. Özkar, "Hydrogen generation from the hydrolysis of sodium borohydride by using water dispersible, hydrogenphosphate-stabilized nickel(0) nanoclusters as catalyst," International Journal of Hydrogen Energy, vol. 32, no. 12, pp. 1707–1715, 2007.

24. J. C. Ingersoll, N. Mani, J. C. Thenmozhiyal, and A. Muthaiah, "Catalytic hydrolysis of sodium borohydride by a novel

nickel-cobalt-boride catalyst," Journal of Power Sources, vol. 173, no. 1, pp. 450–457, 2007.

25. R. Peña-Alonso, A. Sicurelli, E. Callone, G. Carturan, and R. Raj, "A picoscale catalyst for hydrogen generation from $NaBH_4$ for fuel cells," Journal of Power Sources, vol. 165, no. 1, pp. 315–323, 2007.

26. M. Ziabari, V. Mottaghitalab, and A. K. Haghi, "A new approach for optimization of electrospun nanofiber formation process," Korean Journal of Chemical Engineering, vol. 27, no. 1, pp. 340–354, 2010.

27. G. T. Kim, Y. C. Ahn, and J. K. Lee, "Characteristics of Nylon 6 nanofilter for removing ultra-fine particles," Korean Journal of Chemical Engineering, vol. 25, no. 2, pp. 368–372, 2008.

28. F. A. Sheikh, N. A. M. Barakat, M. A. Kanjwal, S. J. Park, D. K. Park, and H. Y. Kim, "Synthesis of poly(vinyl alcohol) (PVA) nanofibers incorporating hydroxyapatite nanoparticles as future implant materials," Macromolecular Research, vol. 18, no. 1, pp. 59–66, 2010.

29. M. A. Kanjwal, N. A. M. Barakat, F. A. Sheikh, W. I. Baek, M. S. Khil, and H. Y. Kim, "Effects of silver content and morphology on the catalytic activity of silver-grafted titanium oxide nanostructure," Fibers and Polymers, vol. 11, no. 5, pp. 700–709, 2010.

30. Y. Chen and H. Kim, "Use of a nickel-boride-silica nanocomposite catalyst prepared by in-situ reduction for hydrogen Production from hydrolysis of sodium borohydride," Fuel Processing Technology, vol. 89, no. 10, pp. 966–972, 2008.

31. F. A. Sheikh, M. A. Kanjwal, H. Y. Kim, and H. Kim, "Fabrication of titanium dioxide nanofibers containing hydroxyapatite nanoparticles," Applied Surface Science, vol. 257, no. 1, pp. 296–301, 2010.

32. F. A. Sheikh, N. A. M. Barakat, M. A. Kanjwal et al., "Electrospun titanium dioxide nanofibers containing hydroxyapatite and silver nanoparticles as future implant materials," Journal of

Materials Science: Materials in Medicine, vol. 21, no. 9, pp. 2551–2559, 2010.

33.   F. A. Sheikh, N. A. M. Barakat, M. A. Kanjwal, S. H. Jeon, H. S. Kang, and H. Y. Kim, "Self synthesize of silver nanoparticles in/on polyurethane nanofibers: nano-biotechnological approach," Journal of Applied Polymer Science, vol. 115, no. 6, pp. 3189–3198, 2010.

34.   M. A. Kanjwal, N. A. M. Barakat, F. A. Sheikh, M. S. Khil, and H. Y. Kim, "Functionalization of electrospun titanium oxide nanofibers with silver nanoparticles: Strongly effective photocatalyst,"International Journal of Applied Ceramic Technology, vol. 7, no. 1, pp. E54–E63, 2010.

35.   JCDPS card no 21-1272.

36.   JCDPS card no 7440-02-0.

4

Simulation and Optimization of Natural Gas Processing and Production Network Consisting of LNG, GTL, and Methanol Facilities

S.A. Al-Sobhi and A. Elkamel

Department of Chemical Engineering, University of Waterloo, Waterloo, Ontario, Canada

ABSTRACT

A considerable growth is expected for the natural gas sector and some believe that it will be the leading primary fuel between

2020 and 2030. Different options are available to utilize the natural gas resource. Three processes namely, LNG, GTL, and methanol considered as the most promising utilization options are addressed in this paper to represent a natural gas processing and production network. The objective of this work is to illustrate the importance of incorporating rigorous simulation models in the decision-making process in the gas processing industry. A steady state simulation is carried out for LNG, GTL, and methanol processed to determine mass and energy balances, operating conditions, and equipment specification. The simulations' flowsheet is beneficial in many aspects. For example, accurate yield values can be obtained. Also, both the capital and operating costs can be estimated. Moreover, the environmental impact can be assessed quantitatively. Then, an optimization model is presented that is able to represent the processing and production network over a wide range of forecasted economic changes. The main feature of this work is the usage of an integrated simulation-optimization framework. Although the modeling, simulation and optimization of a single natural gas system have been addressed previously, the simulation and optimization of enterprise-wide natural gas processing has not been addressed to this extent in the literature. Furthermore, besides considering more than one utilization process, namely LNG, GTL, and methanol, this work addresses the preprocessing units of these utilization processes in a comprehensive manner. The results of the optimization are improved by utilizing data from process simulation. Such data are used to tune the optimization model. An illustrative case study is used to show the applicability of formulated optimization model on natural gas processing and production network and show how accurate representations of the plants are obtained from process simulation. The end result is a more highly optimized and sustainable processing and production network.

INTRODUCTION

Natural gas is the cleanest primary fossil fuel. It produces less CO_2, NO_x, SO_x, and particulates emissions when burned to produce energy compared to other traditional fossil fuels such as, oil and coal (Washington, 1999). In 2012, it was used and consumed globally to generate electricity by 24% relative to the other energy sources, namely, oil, coal, nuclear energy, hydroelectricity, and renewables (BP Statistical Review, 2011). Global concerns and stringent environmental regulations, all motivate looking for alternative fuels and even prioritizing among available ones. Therefore, natural gas with its abundance and relatively cheap prices has a clear advantage and plays an important role in energy mix. A considerable growth is expected for natural gas sector and some believe that it will be the leading fuel between 2020 and 2030 (Economides and Wood, 2009).

There are a number of ways to monetize the natural gas resource. The utilization process depends on the resource location, quantity, quality and so on. Conventionally, pipelines are used to transfer the natural gas to the final consumers. In many cases, this option is not possible and even not practical when the resource is stranded and the transportation distance is long. Other options considered promising as reported by Thomas and Dawe (Thomas, 2003), include: Liquefied natural gas (LNG), compressed natural gas (CNG), gas to solid (GTS), i.e. hydrates, gas to wire (GTW), i.e. electricity, gas to liquids (GTL), and gas to chemical. Moreover, there is a growing recognition of the unconventional gas sources, such as shale gas, coal bed methane (CBM) and deep tight gas, and it is expected that they will contribute significantly in the future (Economides and Wood, 2009).

Many researchers have addressed these utilization options either individually or as a combination. They assessed the utilization option from design, operation, economic, and environment impact perspectives. For example, Khalilpour and Karimi (Khalilpour and Karimi, 2012) considered LNG, CNG, and GTL as monetization options under the uncertainty of natural gas and oil prices. Wood et al., (Wood et al., 2012) presented a review study addressing the available opportunities for GTL industry. Bao et al., (Bao et al., 2010) presented a techno-economic study of GTL process. GTL economic viability depends strongly on oil prices because the GTL products such as low-sulfur gasoline and diesel are oil refinery products competitors. Oil prices of $20–30 per barrel will justify the decision of considering GTL process as option of utilization according to studies done by Economides (Ol and Economides, 2005) and Al-Saadoon (Al-Saadoon, 2007). With current oil price around $100 per barrel (last quarter of 2013), simply the decision of considering GTL process is totally justified. Furthermore, LNG was accounted for 30.5% of global natural gas trade in 2010 (BP Statistical Review, 2011). Methanol that is a primary feedstock for chemical industry is expected to contribute as fuel and energy carrier. Methanol when combined with dimethyl ether (DME) represents excellent fuels. Furthermore, methanol and DME can be blended with gasoline/ diesel and used in internal combustions engines and in electricity generators (Olah et al., 2006). For chemical sector, methanol to olefins and methanol to hydrogen are promising applications (Haid and Koss, 2001). For the abovementioned reasons, the three most attractive utilization processes, namely LNG, GTL, and methanol are considered in this paper to represent the processing and production network.

We present a framework that addresses the simulation, modeling, and optimization of natural gas processing and production network consisting of LNG, GTL, and methanol facilities as shown in Fig. 1. It is worth mentioning that this work addresses both the onshore processing and production and offshore processing is beyond the scope of this research. The upstream processing is a common processing part among all facilities includes the stabilization, acid gas removal, sulfur recovery, dehydration, and natural gas liquids (NGL) separation units. The impact of such consideration is seen when different natural gas feed flowrate and composition are introduced to the upstream processing side and how this affect the yields obtained from downstream side of the network. The objective of this paper is to illustrate and emphasize the importance of incorporating the rigorous simulation models in decision-making process in the gas processing industry. We will show that such incorporation improves the validity of the optimized models that are used for production planning of the gas processing network.

The resulting optimization model is able to represent the processing and production network over a wide range of economic changes as forecasting suggest.

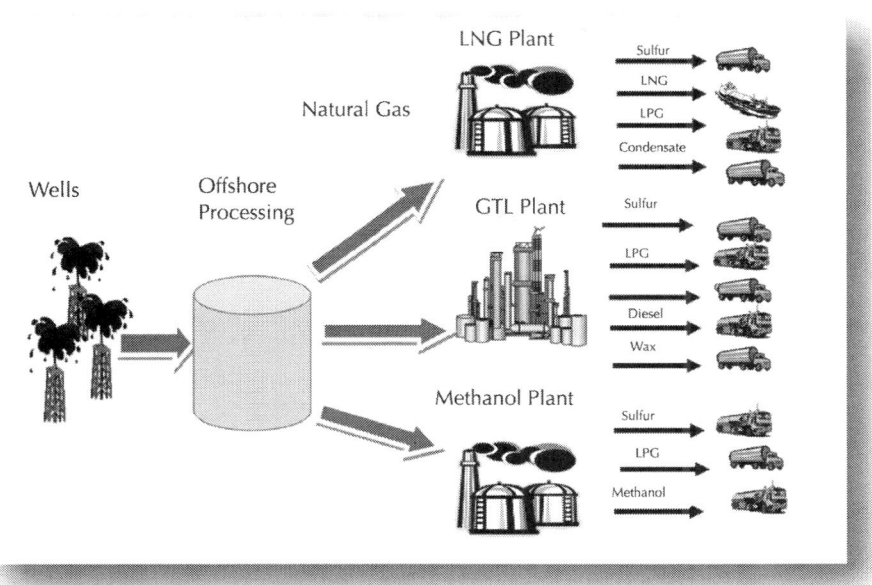

Figure 1: A schematic representation of natural gas processing and production network.

BACKGROUND: PROCESS DESCRIPTION

LNG Process Description

Natural gas is converted physically through compression and liquefaction to LNG. Natural gas is liquefied at −160 °C and 1 atm and this will reduce its volume by 600 times. Fig. 2 shows a typical block flow diagram of an LNG train. Natural gas feedstock is received at inlet receiving unit where the initial gas–liquid separation and metering take place. The hydrocarbon condensate stream is stabilized in the condensate stabilization unit. Acid gas removal unit is designed to reduce the CO_2 and H_2S concentration levels to specification limits. The specifications as reported by Klinkenbijl et al. (Klinkenbijl, 2005) are set to be lower than 50 ppmv for CO_2 to prevent freezing in the main cryogenic heat exchanger and below 3.5 ppmv for H_2S to meet sale gas and pipeline specifications. Sulfur is recovered as by-product. Gas leaving the acid gas removal unit is called sweet gas. This treated gas is passed to dehydration, mercaptan, and mercury removal unit. The dehydration involves water vapor removal to a very low level for example below 0.5 ppmv. Mercury is removed to small level as 0.01 µg per normal cubic meter (Nm3). Heavier hydrocarbons are recovered in the pre-cooling NGL separation unit. They are sent to fractionation unit where ethane, propane, butane, and plant condensate are recovered. The gas rich in mainly methane leaving the gas pre-cooling, NGL separation unit is liquefied by one of approved large scale baseload natural gas liquefaction processes such as pure refrigerant cascade process, propane precooled mixed refrigerant process, propane precooled mixed refrigerant with back end nitrogen expander cycle, or other mixed refrigerant processes (Tusiani and Shearer, 2007).

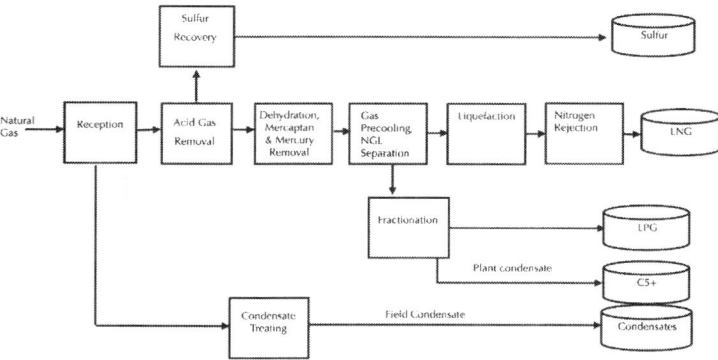

Figure 2: A typical LNG train block flow diagram.

GTL Process Description

Natural gas is converted chemically through Fischer–Tropsch (FT) reaction into liquid fuels (Steynberg and Dry, 2004). Fig. 3 shows a typical block diagram of a GTL train. The upstream processing units such as stabilization, acid gas removal, sulfur recovery, dehydration, and natural gas liquids separation units are similar to those in LNG train. However, the downstream processing units consist mainly of three basic units. These are namely (1) Reforming or synthesis gas production unit where the proper syngas ratio (H_2/CO) is produced. (2) FT synthesis unit where the synthesis fuels (synfuels) are produced (3) Product upgrading and separation unit where the hydro-treating/cracking takes place to obtain the final liquid fuels.

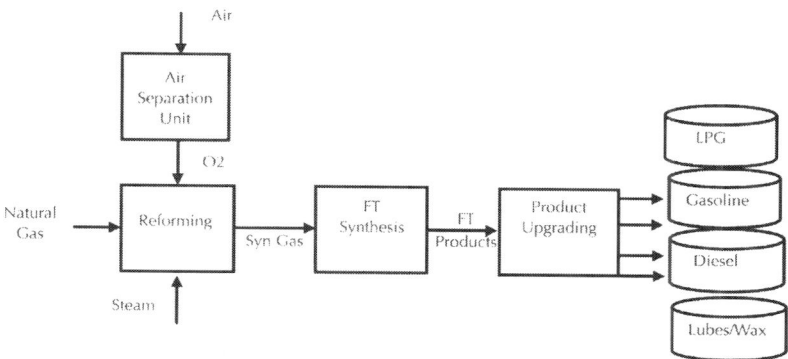

Figure 3: A typical GTL train block flow diagram.

Methanol Process Description

Natural gas is converted chemically through reactions into methanol. Fig. 4 shows a typical block flow diagram of a methanol train. The upstream processing units such as stabilization, acid gas removal, sulfur recovery, dehydration, and natural gas liquids separation units are similar to those in LNG train.

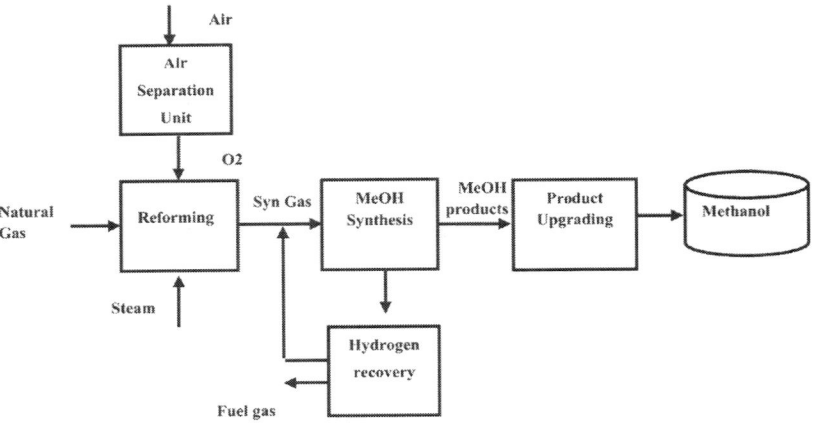

Figure 4: A typical methanol train block flow diagram.

However, the downstream processing units consist mainly of three basic units (Olah et al., 2006). These are namely (1) Reforming or synthesis gas production unit where the proper molar syngas stoichiometric ratio SR = $(H_2–CO_2/CO + CO_2)$ is produced. (2) Methanol synthesis unit where the crude methanol is produced (3) Product upgrading unit where the separation and distillation take place to obtain the final methanol product.

PROBLEM STATEMENT AND SOLUTION STRATEGY

A natural gas processing and production network consisting of LNG, GTL, and methanol facilities is considered. The network involves the processing of natural gas feedstock to produce different set of main products such as LNG, gasoline, diesel, methanol, and by-products such as, sulfur, LPG, and condensate. The network is envisioned as a set of interconnected nodes representing the main processing units of three facilities. These nodes are connected by arcs that represent the material flows within each facility. The utilization options are specified and the problem boundaries are defined, for example, a specific flowrate, composition, and operating conditions. It is worth mentioning that considering a specific number of utilization processes will just represent the base case. Furthermore, any promising processes can be included in the future planning to address process flexibility. For example, hydrogen as a main product from natural gas, methanol to olefin (MTO) or methanol to gasoline (MTG) production process from methanol (Wood et al., 2012 and Haid and Koss, 2001) or even hydrogen from methanol.

The available technologies as shown inTable 1 are then screened. After consulting some published works and books (Olah et al., 2006, Tusiani and Shearer, 2007, Steynberg and Dry, 2004, Robert and John, 1998 and Mokhatab and Poe, 2012) in natural gas processing industry, the most applicable one is selected for steady state simulation. In other words, we fix the topology of the superstructure for steady state simulation. The use of any process simulators (such as CHEMCAD, Aspen Plus, HYSYS, and PRO/II) will be beneficial at this step.

The steady state simulation of the selected process flowsheet is carried out using ASPEN Plus V7.3 version (Aspen Plus). Essentially, the material and energy balances are calculated at this step for each process. Products yield is obtained to be used in the mathematical programming step. After the steady state simulation is converged, equipment sizing is estimated. Then, Aspen process Economic Analyzer is used to estimate both the fixed capital and operating costs of each key unit of selected processes to be used in the mathematical model. A mathematical programming model is formulated as detailed in Section 4 for optimal operation. The yields obtained from the simulation are used as parameters in the optimization step. The modeling software LINGO version 14.0 (LINGO) is used to run the model and obtain the optimal results. The solution strategy is depicted in Fig. 5.

Table 1: Possible technology of industrial processing units

Major processing unit	Possible processes/technologies	Selected process/technology
Condensate stabilization	(a)Flash vaporization (b)Stabilization by fractionation	Stabilization by fractionation
Acid gas removal	(a)Indirect conversion (b)Direct conversion, i.e. (dry bed or liquid phase) (c)Separation technologies, i,e (membrane or cryogenic fractionation)	Direct conversion (liquid phase)
Sulfur recovery	(a)Gas phase-Claus type (b)Liquid-phase	Gas phase Claus type
Dehydration	(a)Liquid desiccant (glycol) (b)Solid desiccant (c)Cooling the gas	Glycol
NGL recovery (extraction)	(a)Refrigeration process (b)Lean oil absorption (c)Solid bed adsorption (d)Membrane separation (e)Twister supersonic	Refrigeration process
NGL fractionation	(a)Direct sequence (b)Indirect sequence	Direct sequence

Liquefaction	(a)Pure-refrigerant cascade	Propane-precooled mixed-refrigerant
	(b)Propane-precooled mixed-refrigerant	
	(c)Propane-precooled mixed-refrigerant, with back-end nitrogen expander-based	
	(d)Nitrogen expander-based	
Reforming (syngas production)	(a)Steam reforming	Auto-thermal reforming
	(b)Adiabatic oxidative reforming	
	(c)Auto-thermal reforming	
FT synthesis	(a)Low temperature FT	Low temperature FT
	(b)High temperature FT	
Methanol synthesis	(a)Quench	Quench
	(b)Steam raising	
	(c)Gas cooled (tubular)	
Product upgrading	(a)Direct sequence distillation	Direct sequence distillation
	(b)Indirect sequence distillation	

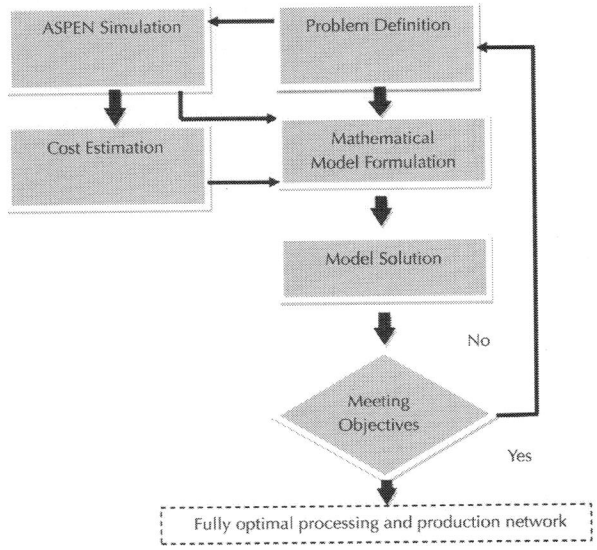

Figure 5: A schematic representations of steps in developing optimal structure of network.

MATHEMATICAL FORMULATION

Modeling the natural gas supply chain is a challenging task. Therefore, researchers try to model the main two components; production and distribution networks separately. This simplifies the formulation, convergence, and increases the level of accuracy significantly. The network includes both upstream processing and downstream production.

After modeling the processing and production network we can answer several tactical and operational questions such as:

- What is the optimal natural gas flowrate to each process?
- What is the optimal production level of each product? Which process can increase the economic portfolio of processing network? How to plan under the price fluctuations?
- Which process is more sustainable; for example, producing

less CO_2 emissions? How to make the processing network more sustainable?

- The effect of gas composition on the overall structure and optimal planning strategy?
- The effect of demand on available capacity and expansion strategy in the processing network?

The gas processing and production network consists of a set of plants or nodes j ∈ J that are of main types LNG, GTL, and methanol. Each node consists of a set of main processing units that are connected in a specified way and this gives the identity to the node in term of consumption of raw materials, production of final products, utility requirements, and environmental impact as shown in Fig. 6. In order to propose an appropriate mathematical model we first define the following sets, indices, parameters, and variables:

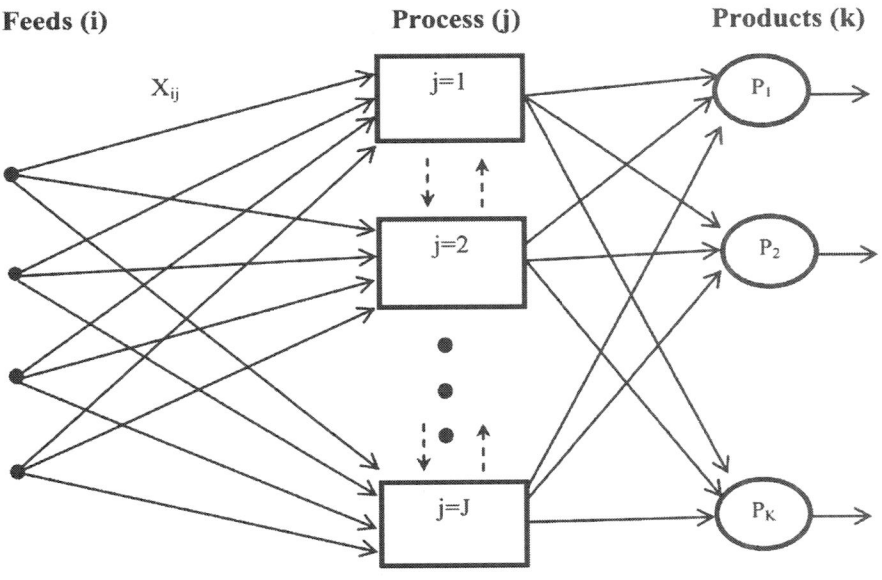

Figure 6: A schematic superstructure of natural gas processing network.

Sets

I = {1, 2, 3, …, I} feedstock which are fed to processing network consisting of different set of processes (nodes)

J = {J1, J2, J3, …} where Ji = {LNG}, {GTL}, or {Methanol} producing different set of products K = {1, 2, 3, …, K} by using different utilities U = {1, 2, 3, …,U}.

Variables

X_{ij} = mass flowrate of natural gas i entering process j

Y_{jk} = mass flowrate of product k from process j

$$Xi = \sum_{j=1}^{J} X_{iJ}$$
= total mass flowrate of natural gas fed to processing network

$$Xj = \sum_{i=1}^{I} X_{ij}$$
= total mass flowrate entering process/node j

Euj = utility requirement per unit feed to unit j for utility u

Objective function is set to maximize profit,

Maximize

$$\sum_{j=1}^{J}\sum_{k=1}^{K} C_{jk}Y_{jk} - \sum_{i=1}^{I}\sum_{j=1}^{J} C_{ij}X_{ij} - \sum_{j=1}^{J} f(Xj).Xj - \sum_{j=1}^{J}\sum_{u=1}^{U} C_{u}g(E_{uj})$$

(1)

The objective function is calculated as the sum of product values minus the cost of raw material, the operating cost, and the cost of utility consumption. The function f(Xj) and g(Euj) are in general nonlinear function with f(X$_j$) indicating the variable production cost with amount produced by unit j and g (E$_{uj}$) indicates the utility consumption function as a function of utility requirement per unit j.

System Constraints

The main constraint covering the processing network is the material balance constraint and will be represented by the yield obtained from the steady state simulation.

Supply Constraint

Total natural gas allowable usage from the field per day

$$\sum_{i=1}^{I} X_{ij} \leq X_i^U$$

(2)

Process Capacity Constraint

Lower and upper capacity for each process

$$X_j^L \leq X_j \leq X_j^U$$

(3)

Demand Constraint

$$Y_k^L \leq Y_k \leq Y_k^U$$

(4)

Material Balance: Fixed Plant Yield

$$Y_k = \sum_{j=1}^{J} Y_{jk}$$

(5)

$$Y_{jk} = \sum_{j=1}^{J} \sum_{i=1}^{I} Y_{ij} X_{ij}$$

(6)

Utility Consumption Constraint

$$E_{uj} \leq E_u^U$$

(7)

Non Negativity Constraint

All the variables are positive.

CASE STUDY

An illustrative case study is presented to show the applicability of the framework presented earlier. The steady state simulation of the processing network was carried out using Aspen Plus as mentioned before. The main key processing units of LNG, GTL, and methanol are considered. A typical natural gas composition (mol%) as shown in Table 2 with a specific flowrate and operating conditions is used in our analysis.

Table 2: Natural gas feed operating conditions and composition (Al-Sobhi et al., 2009)

Flow rate	1500 MSCFD 74,700 kmol/h 1.5 × 10⁶ kg/hr
Temperature	20 °C
Pressure	70 bar
Component	mol%
H_2S	1
CO_2	2
N_2	4
H_2O	0.05
CH_4	83
C_2H_6	5
C_3H8	1.8

i-C_4H_{10}	0.4
n-C_4H_{10}	0.7
i-C_5H_{12}	0.3
n-C_5H_{12}	0.3
n-C_6H_{14}	1.45
Total	100

LNG Process Simulation

Natural gas feedstock with flowrate, operating conditions, and mole composition shown in Table 2, is received at inlet receiving unit. The steady state simulation of LNG process was carried out using ASPEN Plus as shown in Fig. 7. The primary gas/condensate separation takes place and condensate C5+ product is stabilized. In general, cubic equation of state, for example, Peng Robinson or Soave–Redlich–Kwong equation (SRK) is appropriate and recommended by ASPEN for gas processing. The Peng Robinson thermodynamic property method is considered and selected as global thermodynamic property method. However, NRTL was selected as thermodynamic property method for amine system. The residual sour gas is then fed to Acid gas removal unit where H_2S and CO_2 are removed using the mixed amine solution. The composition of mixed amine solution is (15 wt% DEA, 30 wt% MDEA, 45 wt% water). The CO_2 and H_2S mole fraction in the sweet gas stream are 7.9 E-5, and 1.8 E-25, respectively. The elemental sulfur is recovered from the H_2S rich stream by the conventional Clause process (Robert and John, 1998). The sweet gas is fed to dehydration unit, where TEG glycol is used to remove the water content to 9.0 E-05 as water mole fraction. The sweet dry gas stream is fed to NGL unit where the heavy hydrocarbons and LPG are separated from the gas that is mainly methane. The methane gas is liquefied to −160 °C using a mixed refrigerant (MR). The optimized MR composition is obtained from Alabdulkarem et al. A detailed discussion about cryogenic processes can be found in Venkatarathnam (2008). Then, the nitrogen is rejected to obtain the desired heating value of LNG product. The products mass flowrates and yields are shown in Table 3.

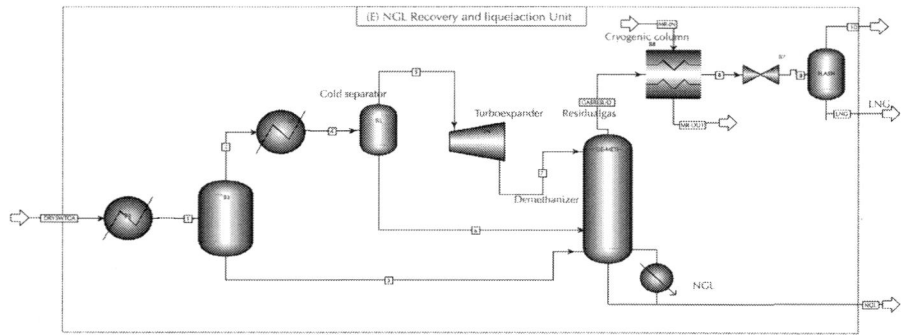

Figure 7: ASPEN Plus flowsheet of the NGL and liquefaction unit.

Table 3: Products yield from LNG, GTL, and methanol processes

Product (kg/hr)	NG1	NG2	NG3	Yield[a]	Min. demand	Max. demand	LINGO Model's output
Sulfur1	22,900	–	–	0.015	20,000	30,000	20,000
LNG1	1,044,157	–	–	0.69	1,000,000	1,200,000	1,090,000
Ethane1	62,800	–	–	0.04	60,000	70,000	60,000
LPG1	75,200	–	–	0.05	70,000	80,000	80,000
Plant Condensate1	12,700	–	–	0.01	10,000	15,000	10,000
Field Condensate1	64,900	–	–	0.043	60,000	70,000	60,000
Losses[b]1	232,343	–	–	0.152	230,000	235,000	230,000
Sulfur2	–	22,900	–	0.015	20,000	30,000	21,150
Eathane2	–	62,800	–	0.04	60,000	70,000	70,000
LPG2	–	114,000	–	0.075	110,000	120,000	120,000
Gasoline2	–	227,911	–	0.15	220,000	230,000	230,000
Diesel2	–	174,730	–	0.11	170,000	180,000	180,000
Wax2	–	99,900	–	0.07	95,000	100,000	100,000
Plant Condensate2	–	12,700	–	0.01	10,000	15,000	15,000

Field Condensate2	–	64,900	–	0.043	60,000	70,000	70,000
Losses[b]2	–	735,159	–	0.48	733,000	735,200	733,000
Sulfur3	–	–	22,900	0.015	20,000	30,000	30,000
Eathane3	–	–	62,800	0.04	60,000	70,000	70,000
LPG3	–	–	75,200	0.05	70,000	80,000	80,000
Methanol	–	–	870,000	0.574	850,000	900,000	900,000
Field Condensate3	–	–	64,900	0.043	60,000	70,000	70,000
Losses[b]3	–	–	419,200	0.27	400,000	410,000	400,000
Operating cost($/yr)	6.74E+5	7.22E+5	6.84E+5	–	–	–	–
Available NG supply(kg/hr)	1,515,000	1,515,000	1,515,000	–	–	–	–

a Yield defined as product flowrate divided by feedstock flowrate.

b Losses represent other byproducts such as CO_2, N_2, water, etc.

GTL Process

The treated methane gas is preheated, mixed with steam and pure oxygen, and fed to Auto-thermal reactor (ATR). The ATR converts the natural gas that is mainly methane, steam, and pure oxygen into a syngas (H_2/CO). The ATR reaction scheme is complex, but overall reaction is represented by Steynberg and Dry (2004):

$$CH_4 + 3/2O_2 \rightarrow CO + 2H_2O - \Delta H^{\circ}_{298} = +519 \text{ KJ/mole}$$

(8)

$$CH_4 + H_2O \leftrightarrow CO + 3H_2 - \Delta H^{\circ}_{298} = -206 \text{ KJ/mole}$$

(9)

$$CO + H_2O \leftrightarrow CO_2 + H_2 - \Delta H^{\circ}_{298} = +41 \text{ KJ/mole}$$

(10)

The ATR is modeled as Equilibrium reactor in ASPEN Plus as shown in Fig. 8. For the given natural gas flowrate, the steam to CH_4 is set to be 0.6 as the operating ratio. This very low ratio around 0.6, rather than the previously used high ratio of 1.5–2.0, becomes the state-of-the-art syngas ratio for FT application in modern plants in Europe and Middle East (Steynberg and Dry, 2004).

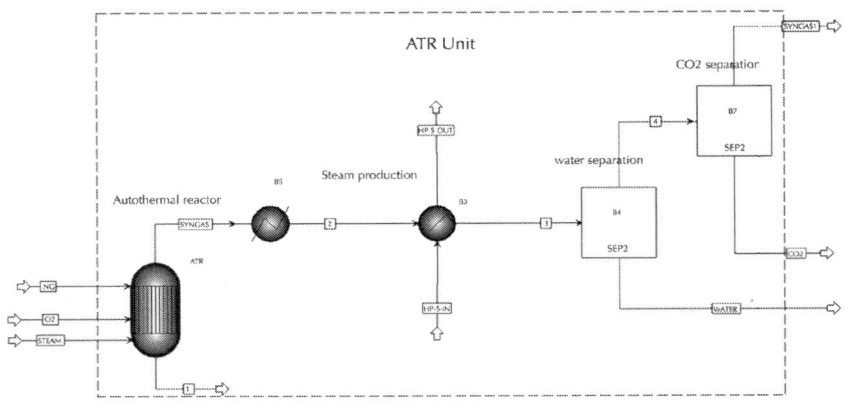

Figure 8: ASPEN Plus flowsheet of the ATR unit.

Now, For the O_2 flowrate sensitivity analysis was performed by varying the O_2 flowrate to obtain the desired H_2/CO ratio as seen in Fig. 9. The syngas H_2/CO ratio of 2 is achieved by using the ATR with 70,000–72,500 kmol/h as O_2 flowrate.

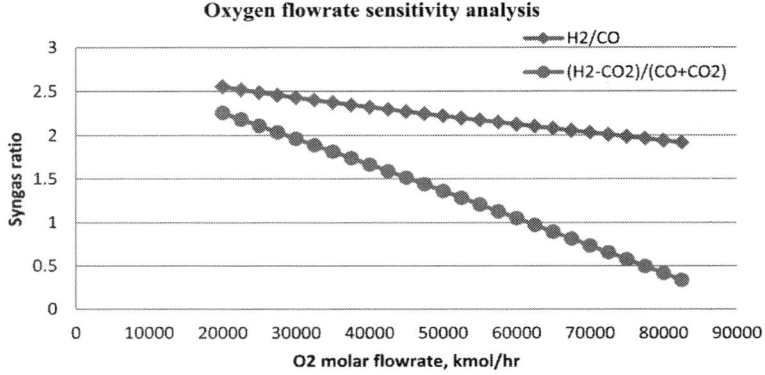

Figure 9: Sensitivity analysis by varying O_2 flowrate.

The high syngas temperature is utilized by producing high pressure steam. The syngas is fed to slurry phase FT reactor which is modeled as Yield reactor. The FT reactor operating conditions are 240 °C and 20 bar. The FT synthesis

$$nCO + 2nH_2 \rightarrow \{-CH_2-\}n + nH_2O$$

$$(11)$$

This study is considered as a low-temperature slurry-phase Fischer–Tropsch reactor with cobalt catalyst. The product distribution follows the chain growth probability function known as Anderson–Schulz–Flory (ASF) distribution (Steynberg and Dry, 2004).

$$\frac{W_n}{n} = (1 - \alpha)^2 \alpha^{n-1}$$

$$(12)$$

where Wn is the mass fraction of the hydrocarbon molecular with carbon number n and   is the chain growth probability. An alpha value of 0.9 is selected to find the mass fraction of the hydrocarbons as shown in Fig. 10. FT reactor is assumed to produce only paraffine (no olefin). The hydro-treating/cracking of the waxes takes place to obtain the final desired products normally LPG, synthetic gasoline and diesel. The products mass flowrates and yields are shown in Table 3.

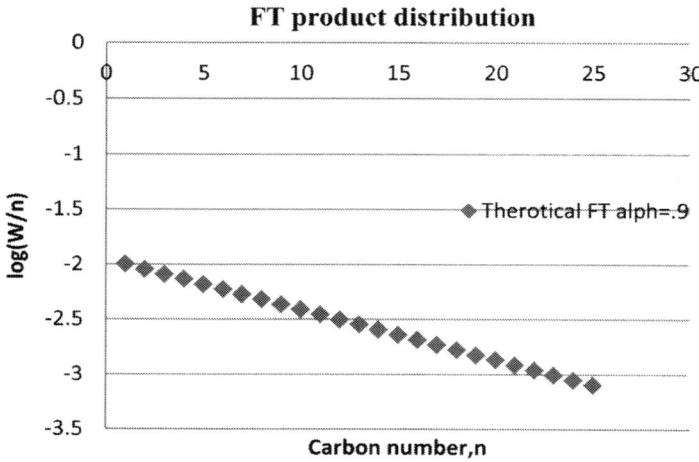

Figure 10: FT product wt% distribution for = 0.9.

Methanol Process Simulation

The treated methane gas is preheated, mixed with steam and pure oxygen, and fed to Auto-thermal reactor (ATR) similar to the one used in GTL process. The ATR converts the natural gas that is mainly methane, steam, and pure oxygen into proper syngas stoichiometric ratio SR = (H_2–CO_2/CO + CO_2). Now, one of the advantages of using the rigorous simulation is highlighted by using the sensitivity analysis modeling tool in Aspen plus. For the O_2 flowrate, sensitivity analysis was performed by varying the O_2 flowrate to obtain the desired H_2/CO and SR ratio of 2 for GTL and methanol applications. The syngas (H_2/CO) and SR ratio of 2 is achieved by using the ATR with 72,500 and 27,500 kmol/h as O_2 flowrate for GTL and methanol applications, respectively as seen in Fig. 9. The syngas with SR of 2 is fed to methanol synthesis reactor which is modeled as equilibrium reactor. The methanol synthesis typically consists of mainly three reactions, where the two reactions in equations (13) and (14) are exothermic, with heat of reaction equal to −21.7 kcal/mol and −9.8 kcal/mol, respectively:

$$CO + 2H_2 \leftrightarrow CH_3OH - \Delta H^\circ_{298} = -21.7 \text{ kcal/mole} \tag{13}$$

$$CO_2 + 3H_2 \leftrightarrow CH_3OH + H_2O - \Delta H^\circ_{298} = -9.8 \text{ kcal/mole} \tag{14}$$

$$CO + H_2O \leftrightarrow CO_2 + H_2 - \Delta H^\circ_{298} = 11.9 \text{ kcal/mole} \tag{15}$$

First of all, since the synthesis reactions are highly exothermic, heat released in the synthesis reaction should be either recovered for power generation or absorbed by cooling water to obtain an isothermal operation. Equation (15) describes the endothermic reverse water gas shift reaction (RWGSR) that also occurs during methanol synthesis, producing CO which can be further react with hydrogen to produce methanol. The methanol reactor operating pressures range from 50 to 100 atm and temperatures of 200–300 °C. This low pressure route is the basis for most methanol production processes. Crude methanol leaving the reactor contains water and some impurities depending on feed gas composition, reaction conditions, and type and lifetime of the catalyst, such as dissolved gases (methane, CO, CO_2), higher alcohols (ethanol, propanol, butanol) and long-chain hydrocarbons. Methanol is available in three grades of purity: (1) fuel grade, (2) "A" grade, used as a solvent, (3) "AA" grade or chemical grade with highest purity with 99.85% methanol content (Olah et al., 2006). The distillation systems using one or more distillation columns will be used to purify the methanol product. The products mass flowrates and yields are shown in Table 3.

RESULTS AND DISCUSSIONS

The model has been solved for different scenarios.

Scenario 1: Network Optimization (Base Case)

After carrying out the steady state simulation of LNG, GTL, and methanol processes, the products mass flowrate in kg/hr, yield, and max/min demand of LNG, GTL, and methanol are found for the base case conditions as shown in Table 1 and economic data as shown in Table 4. The LP model has 25 variables and 81 constraints. It has been solved in LINGO 14.0 version. The optimal values of products flowrate are tabulated in the last column in Table 3. It was found' that $218,788 is the optimal hourly profit as defined by the objective function. Furthermore, 1,550,000, 1,550,000 and 1,550,000 kg/hr were the optimal natural gas feedstock flowrate to LNG, GTL, and methanol facilities, receptively.

Table 4: Economic data

Natural gas	$ 4.4 per MMBtu
Sulfur	$ 200 per tonne
Ethane	$ 10 per MMBtu
LNG	$ 7 per thousand cubic feet
LPG	$ 2.5 per gallon
Plant Condensate	$ 15 per MMBtu
Field Condensate	$ 12 per MMBtu
Gasoline	$ 2.8 per gallon
Diesel	$ 3 per gallon
Wax	$ 2 per gallon
Methanol	$ 500 per tonne

Scenario 2: Natural Gas Feedstock Flowrate Incremental

In this scenario, we consider the increase in natural gas feedstock flowrate by running the simulation for higher flowrate from 1500 to

1800 MMSCFD fed to the stabilization unit. After 1800 MMSCFD, we started to get some converge issues.

So, 1800 MMSCFD is the maximum flowrate to be fed to each processing facility.

Scenario 3: Natural Gas Feedstock and Product Prices Incremental

In this scenario, we consider the increase in both the natural gas feedstock and network products prices. We solve the model for 50% incremental from base case prices.

The optimal values of products flowrate are tabulated in the last column in Table 3. It was found' that $1,489,219 is the optimal hourly profit as defined by the objective function. Furthermore, 1,550,000, 1,550,000 and 1,550,000 kg/hr were the optimal natural gas feedstock flowrate to LNG, GTL, and methanol facilities, receptively.

We can observe from Table 5 that we can maximize the network's profit by processing more natural gas feedstock and intuitively when the products selling price increases by 50% from base case prices. Also, the model solves for lower values for all losses streams because they do not contribute to the processing network's profit. The losses stream from the GTL process is the highest where CO_2, and water are produced in large amounts as wasted products. Thus, further consideration of capturing the CO_2 and utilizing it within the network and incorporating waste water management will improve the performance significantly.

Table 5: Results comparison of three scenarios

	Base case	Scenario 2	Scenario 3
Sulfur	71,150	60,000	60,000
LNG	1,090,000	1,245,000	1,245,000
Ethane	200,000	180,000	180,000
LPG	280,000	420,000	420,000

Plant condensate	25,000	22,000	22,000
Field condensate	200,000	175,000	175,000
Losses**1	230,000	300,000	300,000
Gasoline2	230,000	220,000	220,000
Diesel2	180,000	222,900	222,900
Wax2	100,000	90,000	90,000
Losses**2	733,000	950,000	950,000
Methanol	900,000	1,002,500	1,002,500
Losses**3	400,000	500,000	500,000
Profit ($/hr)	218,788	155,946	1,489,219

CONCLUSIONS AND FUTURE WORK

In this paper, a framework for analysis and optimizing the natural gas network has been presented. The natural gas processing and production network has been synthesized to include LNG, GTL, and methanol facilities. The ASPEN Plus v7.3 process simulator showed to be a powerful package to simulate the key processing units of the processing network. Sensitivity analysis modeling tool allowed us to vary the O_2 flowrate to find the proper syngas and SR ratios. An LP model was formulated to help us optimize the processing network. The products yield obtained from the simulation served as initial values for model's variables. The optimal flowrate values for both network's feedstock and products that maximize the profits were found.

We can maximize the network's profit by processing more natural gas feedstock and intuitively when the products selling price increases. Also, the model solves for lower values for all losses streams because they do not contribute to the processing network's profit. The losses stream from the GTL process has the highest flowrate value where CO_2, and water are produced in large amounts as wasted products. Thus, further consideration of capturing the CO_2 and utilizing it within the network and incorporating waste water management will improve the performance significantly. Next, the multi-period mixed integer modeling and multi-objective optimization of natural gas processing network will be addressed. A multi-objective decision problem, where trade-offs between economic, environmental, and safety concerns need to be resolved. The variation in natural gas composition and in particle the H_2S content incremental needed to be addressed.

ACKNOWLEDGMENTS

The authors would like to acknowledge the financial support from Qatar University as a form of scholarship.

REFERENCES

1.   Al-Saadoon, F.T., 2007. Economics of GTL Plants, pp. 3e5.
2.   Al-Sobhi, S., Alfadala, H., El-Halwagi, M.M., Simulation, Energy, 2009. Integration of a liquefied natural gas (LNG) plant. In: Alfadala, H., Reklaitis, G., Elhalwagi, M.M. (Eds.), Adv. Gas Process. 1st Int. Gas Process. Symp. Elsevier, pp. 131e135.
3.   Alabdulkarem, A., Mortazavi, A., Hwang, Y., Radermacher, R., Rogers, P., 2011. Optimization of propane pre-cooled mixed refrigerant LNG plant. Appl. Therm. Eng. 31, 1091e1098.
4.   Aspen Plus, V7.3, Aspen Technology, Inc. www.aspentech. com. Bao, B., El-halwagi, M.M., Elbashir, N.O., 2010. Simulation, integration, and economic analysis of gas-to-liquid processes. Fuel Process. Technol. 91, 703e713. bp, BP Statistical Review, June, 2011.
5.   Economides, M.J., Wood, D.A., 2009. The state of natural gas. J. Nat. Gas Sci. Eng. 1, 1e13.
6.   Haid, J., Koss, U., 2001. Lurgi's mega-methanol technology opens the door for a new era in down-stream applications. Stud. Surf. Sci. Catal. 399e404.

7. Khalilpour, R., Karimi, I.A., 2012. Evaluation of utilization alternatives for stranded natural gas. Energy 40, 317e328.

8. Klinkenbijl, A.J., 2005. Best Practice for Deep Treating Sour Natural Gases (to LNG and GTL).

9. LINGO,V14, LINDO Systems, Inc. www.lindo.com.

10. Mokhatab, S., Poe, W.A., 2012. Handbook of Natural Gas Transmission and Processing. Gulf Professional Publishing, Waltham, MA, USA.

11. Ol, V.N.O., Economides, M.J., 2005. The economics of gas to liquids compared to liquefied natural gas. 8, 136e140.

12. Olah, G.A., Goeppert, A., Prakash, G.K.S., 2006. Beyond Oil and Gas: the Methanol Economy. WILEY-VCH Verlag GmbH & Co. KGaA, Weinheim.

13. Robert, M., John, M., 1998. Gas conditioning and Processing. In: Gas Treating and Liquid Sweeting, Campbell Petroleum Series, vol. 4. Norman, Oklahoma, USA.

14. Steynberg, A., Dry, M., 2004. Fischer-Tropsch Technology. Elsevier B.V.

15. Thomas, S., 2003. Review of ways to transport natural gas energy from countries which do not need the gas for domestic use. Energy 28, 1461e1477.

16. Tusiani, M., Shearer, G., 2007. LNG, a Nontechnical Guide. PennWell Corporation, Tulsa, Oklahoma, USA.

17. Venkatarathnam, G., 2008. Cryogenic Mixed Refrigerant Processes. Springer-Verlag, New York, LLC.

18. Washington, E., 1999. Natural Gas 1998 Issues and Trends, 0560.

19. Wood, D.A., Nwaoha, C., Towler, B.F., 2012. Gas-to-liquids (GTL): a review of an industry offering several routes for monetizing natural gas. J. Nat. Gas Sci. Eng. 9, 196e208.

Filtration of Gases at High Pressures: Permeation Behavior of Fiber-based Media Used for Natural Gas Cleaning

Murilo D.M. Innocentini[a], Eduardo H. Tanabe[b],
Monica L. Aguiar[b], and José R. Coury[b]

[a]Course of Chemical Engineering, University of Ribeirão Preto, 14096-900 Ribeirão Preto, SP, Brazil

[b]Chemical Engineering Department, Federal University of São Carlos, 13565-905 São Carlos, SP, Brazil

ABSTRACT

The present work investigated the permeation behavior at high-pressure of four commercial-grade media used for natural gas filtration. Samples of nonwoven fabrics (cellulose, polypropylene and polyethylene) and of a stainless steel cloth (dutch twilled pattern) were tested under airflow at room temperature, face velocities from 0.01 to 0.21 m/s and absolute pressures from 93 to 693 kPa. Permeability coefficients k_1 and k_2 based on Forchheimer's equation were fitted from experimental data and correlated to other physical properties, such as fiber diameter, medium thickness and packing density. The ranges found for k_1 (5.74–80.96×10^{-12} m^2) and k_2 (1.32–19.86×10^{-7} m) are typical of fiber media used in filtrations applications. The increase of air pressure in the tested range did not induce deformations in the medium structure, but resulted in proportionally higher pressure drop levels due to the increase of gas density. Based on the experimental results with airflow, the pressure drop was simulated for flow of natural gas at 25 °C and pressures up to 5000 kPa. Considering the combined influences of thickness and permeability coefficients of each filtering medium, pressure drop was found to decrease in the following order: polypropylene>metallic cloth>cellulose>polyester.

INTRODUCTION

Natural gas is a primary energy source with worldwide use in residential, commercial, industrial and power generation segments. The market of natural gas has been fast growing in recent years, especially because of the more stringent legislation on emissions of pollutants into the environment and the improvements in technology for extraction, refining and transportation, which make this fuel and feedstock more available and with more quality to the end user (Cornot-Gandolphe et al., 2003, Victor et al., 2006, DeShazo et al., 2007 and Najibi et al., 2009; Weijermars, 2010). As an example, in 2010, about 45% of the whole Brazilian energetic matrix was based on renewable fuels and other 9.3% represented the contribution of natural gas; in 1990, this contribution was less than 3% (ANEEL, 2008 and IEA, 2010).

Despite the fossil origin, natural gas is considered one of the cleanest fuels, with a molar composition based mostly on low-chain alkanes (methane: 85–91%, ethane: 4–9%, propane: 0.5–2%, others: 1–2%) and with combustion products virtually free of pollutants, such as sulfur oxides and particulate matter. Other beneficial features are the rapid dispersion into the atmosphere in case of leaks, low levels of odors and contaminants and higher heating value than other fossil fuels, such as coal and oil (Guo and Ghalambor, 2005, Kidnay and Parrish, 2006, Speight, 2007 and Wang, 2009).

Several technologies are available to transport natural gas from the production fields to consuming markets, including PNG (pipeline natural gas), LNG (liquefied natural gas), CNG (compressed natural gas) and NGH (natural gas hydrate) (Speight, 2007 and Wang, 2009). Pipeline transportation still dominates the international gas trade, as exemplified by the Bolivia–Brazil pipeline (GASBOL), the longest transmission line in South America, with a total length of 3150 km, nominal diameter from 410 to 810 mm, 16 boosting stations, 41 delivery stations, maximum pressure of 9.8 MPa and a nominal capacity of 30×10^6 m³/day (20 °C, 1 atm). The benefits of natural gas pipeline operations at such higher pressures include the ability to transmit larger volumes of gas through a given size of pipeline, lower transmission losses due to friction and the capability to transmit gas over longer distances without additional boosting stations (Wu, 1998, Mokhatab et al., 2006, Tabkhi, 2007 and Woldeyohannes and Majid, 2011).

An important concern related to the flow of natural gas over long pipeline extensions is the formation of suspended particulate matter, known as black powder. This residue usually results from chemical and microbial reactions on the pipe internal surface due to the presence of moisture and acidic contaminants such as CO_2 and H_2S in natural gas (Baldwin, 1998, Hernandez-Rodriguez et al., 2007, Tsochatzidis and Maroulis, 2007, Trifilieff and Wines, 2009, Azadi et al., 2011, Saremi and Kazemi, 2011 and Yamada et al., 2011). Under dry conditions, black powder can take the form of a very fine powder or solid sediment, while under wet conditions it usually appears as a tar-like substance. Chemical composition includes several forms of iron sulfide and iron oxide, although quartz, iron and calcium carbonates, liquid hydrocarbons and other impurities have also been reported (Baldwin, 1998). Iron sulfide is not easily filtered out of the flow stream and is pyrophoric under some conditions.

The presence of black powder has negative economic implications in natural gas pipeline networks, as it generates wear and reduced efficiency in compressors and clogging of instrumentation and valves. When adhered to the pipe surface, black powder also increases pipe roughness and gas friction, boosting the power required in compressor stations to keep gas flowing properly to destinations. It is estimated that an equivalent of 3–5% of the gas transported is consumed by the compressors in order to compensate for the lost pressure of the gas (Tabkhi, 2007 and Woldeyohannes and Majid, 2011).

Filtration is still the most common method for the removal of dispersed black powder before it enters a compressor, station or processing plant. There are several cleaning devices, such as the basket and cartridge filters, which are usually supplied as a part of the compressor design package when the unit or piping is installed (Baldwin, 1998, Tsochatzidis and Maroulis, 2007, Azadi et al., 2011 and Saremi and Kazemi, 2011). Ideally, these filters should be able to handle suspended wet and dry impurities, collect iron sulfide sub-micron sized particles, trap significant volume of liquids and be cleaned by an automatic backflushed method, which minimizes heavy cake formation, prevents increased levels of flow resistance and prolongs filter life. Other desired filter requirements are the simplicity and safety of cartridge or basket replacement, without compromising the pressure boundary during normal operation.

A good filtering element for natural gas must withstand the high-pressure flow and also provide high collection efficiency with low pressure loss. Commonly used filter media are produced with metallic screens, borosilicate fiberglass, pleated cellulose and felt polyester. Metallic screens can be recovered and reused after cleaning, while media based on polymeric materials are usually disposable (Tanabe et al., 2010, Tanabe et al., 2011a and Tanabe et al., 2011b).

An important aspect concerning natural gas filtration is that it usually occurs at high pressures ($\leq$10 MPa), which implies in a range of fluid properties, cake-medium and fluid-medium interactions that may be very different from those that typically take place during atmospheric aerosol filtration ($\approx$0.1 MPa). At high pressures, flow is negatively affected not only by the increase of gas density and viscosity, but also by any kind of deformation in the filtering medium structure that results in restrictions in the flow path. The extent of these changes on permeation depends on the original morphological and constructive features of the filtering medium, such as thickness, porosity, pore size and fiber size and on their behavior when subjected to elevated pressures.

Due to a number of reasons, much of the empirical knowledge on natural gas filtration acquired in the field is neither sufficient nor reliable for a scientific analysis of the effect of pressure on filtering media, and thus on their permeation and purification performances. The variability in natural gas stream features (including composition and load of black powder and other impurities) along the pipeline extension is included among the reasons.

In this context, the present study is part of a Brazilian research project that aims to investigate systematically the influence of high-pressure flow on the permeation and filtration behavior of different filtering media available in the natural gas market. In particular, this paper deals with the permeation behavior of four different fiber-based media, through tests with airflow under absolute pressures from 93 to 693 kPa and fluid velocities comparable to those that occur during filtration of natural gas in compressor stations. Permeability coefficients were retrieved from airflow experiments and correlated to the applied pressure levels and media features. The acquired permeability data were also helpful in simulations that predicted and compared the pressure drop of filtering media for natural gas flow up to 5000 kPa.

BACKGROUND

For fluid flow purposes, permeability can be broadly defined as a macroscopic measure of how easily a fluid driven by a pressure gradient flows through the voids of a porous medium. In practice, permeability is commonly expressed in the literature through proportionality coefficients of empirical models that allow the quantification of the pressure drop (P) resulting from the percolation of a fluid through the medium with a given volumetric flow rate (Q) or velocity (vs). Permeability is important to filter design, as it ultimately determines the type and power of pumps, blowers and compressors used to force the fluid through the media.

Forchheimer's equation is an example of empirical relationship that describes with reasonable accuracy the flow resistance through porous media for a wide range of fluid velocities. In the differential form, Forchheimer's equation can be expressed as (Innocentini et al., 2005)

$$-\frac{dP}{dx} = \frac{\mu}{k_1} v_s + \frac{\rho}{k_2} v_s^2$$

(1)

in which dP/dx is the pressure gradient along the flow direction x, while μ and   are, respectively, the absolute viscosity and the density of the fluid. The superficial or face velocity v_s is the volumetric flow rate (Q) divided by the surface area of the filter exposed to flow (A). The parameters k_1 and k_2 are known as Darcian and non-Darcian permeability coefficients or constants, respectively, in reference to Darcy's law, which establishes a linear dependence between pressure drop and fluid velocity.

The fluid velocity (vs) through the filter can also be calculated with basis on the mass flow rate (w), fluid density (ρ) and area exposed to flow (A):

$$v_s = \frac{w}{\rho A}$$

(2)

For a constant permeation capacity (w=constant), the integration of Eq. (1) along the filter thickness element (dx) must take into account changes of gas density ( ) and of filter area (A) on face velocity (vs). Gases and vapors experience a volumetric expansion during the passage through the filter that is proportional to the reduction in absolute gas pressure. Since the pressure drop level through the filter elements is relatively low compared to the absolute gas pressure at which filtration occurs, such reduction in gas density can be reasonably represented in permeability modeling with the assumption that the ideal gas law is valid.

On the other hand, the face area (A) depends on the filter geometry. Two common filter geometries (flat panels and hollow cylinders), their respective flow orientations (perpendicular, radially inward and radially outward) and the important medium dimensions for integration of Eq. (1) are schematized in Fig. 1. For radial flow through hollow cylinders, the face area A increases with the radius ($A=\varpi DH$), while it is constant for cross flow through flat membranes.

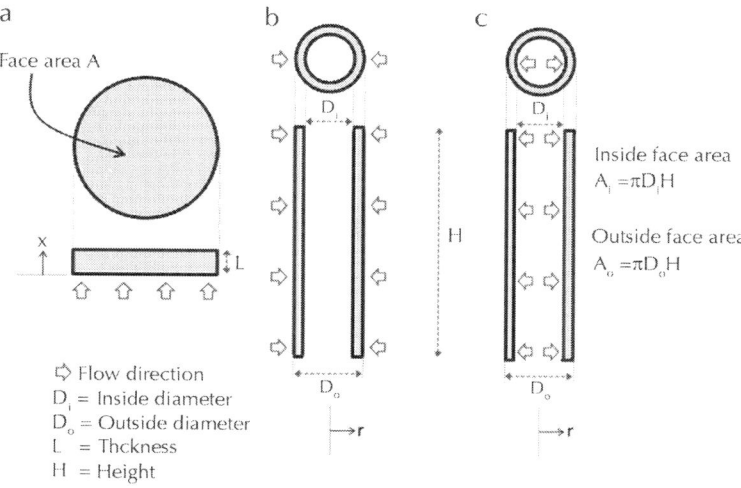

Figure 1: Typical geometries and flow orientation for filter elements; (a) cross flow through flat membrane; (b) radially inward flow through hollow cylinder; (c) radially outward flow through hollow cylinder.

The resulting integrated Forchheimer's equations for rectangular and cylindrical coordinates for compressible flow (variable) are, respectively, given by

–Cross flow through flat membranes:

$$\frac{\Delta P}{L} = \frac{\mu_o}{k_1} v_{so} + \frac{\rho_o}{k_2} v_{so}^2$$

(3)

–Radial flow through hollow cylinders:

$$\frac{\Delta P}{(D_o/2)} = \frac{\mu_o}{k_1} \mathrm{Ln}\left(\frac{D_o}{D_i}\right) v_{so} + \frac{\rho_o}{k_2}\left(\frac{D_o-D_i}{D_i}\right) v_{so}^2$$

(4)

The pressure drop parameter P is calculated by (Innocentini et al., 2005)

$$\Delta P = \frac{P_i^2 - P_o^2}{2P_o}$$

(5)

In Eqs. (3), (4) and (5), L is the medium thickness (for flat membranes) while D_i and D_o are the inside and outside diameters, respectively (for cylindrical filters). P_i and P_o are the absolute gas pressures before and after the filter, respectively. The parameters o, μ_o and v_{so} are, respectively, the density, viscosity and face velocity of the gas based on P_o. Eqs. (3) and (4) can also be deduced with basis on the absolute pressure P_i. For situations in which gas decompression across the filter is unimportant, Eq. (5) is simplified to

$$\Delta P = P_i - P_o$$

(6)

From a physical perspective, Forchheimer's equation states that the loss of pressure energy accompanying the fluid flow is the sum of two contributions that coexist but that change in importance as the fluid velocity increases. There is an agreement in the literature that at low fluid velocities, energy is predominantly dissipated from the fluid to the pore walls by simple viscous action (fluid friction), as it occurs in laminar flow in straight channels (Scheidegger, 1974, Noman and Kalam, 1990, Seguin et al., 1998a, Seguin et al., 1998b, Innocentini et al., 2005 and Hlushkou and Tallarek,

2006). This source of energy (pressure) loss is represented by the linear term in velocity of Eqs. (3) and (4). With the increase of velocity, the curvatures, contractions and enlargements of the flow channels, typical of the pore morphology, generate secondary flow patterns and distortions in the pressure and velocity fields in a microscopic scale that intensify energy dissipation by fluid friction. This increased dissipation leads to a non-linear increase in pressure drop, represented by the quadratic term in velocity of Eqs. (3) and (4). It is worth mentioning that this non-linearity, also known as inertial effect, occurs still within the laminar flow regime and is not necessarily related to turbulence (Scheidegger, 1974, Noman and Kalam, 1990, Seguin et al., 1998a,Seguin et al., 1998b and Hlushkou and Tallarek, 2006).

The permeability coefficients k_1 and k_2 are dimensionally distinct: k_1 is expressed in dimensions of square length (m²), while k_2 is expressed in dimensions of length (m). Both parameters are considered to be only dependent of medium properties and therefore should remain constant even with changes in fluid and flow features. On the other hand, modifications in the flow path caused by temperature effects (such as thermal expansion) or by pressure effects (compaction, stretching or deformation of fibers and granules) may result in changes in k_1 and k_2 values (Innocentini et al., 2005).

In order to assess the flow regime and the validity of empirical permeability correlations for fibrous materials, the fiber Reynolds number (*Ref*) and the Forchheimer number (*Fo*) can be calculated, respectively, by (Innocentini et al., 2005, Hutten, 2007 and Wang, 2007)

$$Re_f = \frac{\rho_o v_{so} d_f}{\varepsilon \mu_o}$$

$$\text{(7)}$$

$$Fo = \frac{\rho_o v_{so}(k_1/k_2)}{\mu_o}$$

$$\text{(8)}$$

The Forchheimer number can be understood as an analog of the fiber Reynolds number, with the characteristic length df being replaced by the ratio k_1/k_2, which also has length dimensions. Fo is related to the linearity in the pressure drop curve in the same way that Ref is related to the laminarity of flow. ForFo 1, inertial effects are negligible and Eq. (3) reduces to Darcy's law (Innocentini et al., 2005):

$$\left(\frac{\Delta P}{L}\right)_{viscous} = \frac{\mu_o}{k_1} v_{so}$$

(9)

On the other hand, when Fo 1, viscous effects can be disregarded and the pressure drop can be reasonably estimated through the inertial-quadratic term:

$$\left(\frac{\Delta P}{L}\right)_{inertial} = \frac{\rho_o}{k_2} v_{so}^2$$

(10)

For any other intermediary flow condition both Eqs. (9) and (10) are relevant and the complete Forchheimer's equation (3) should be used to estimate the total pressure drop through the filter:

$$\left(\frac{\Delta P}{L}\right)_{total} = \left(\frac{\Delta P}{L}\right)_{viscous} + \left(\frac{\Delta P}{L}\right)_{inertial}$$

(11)

The percentage contributions of viscous and inertial pressure drops can be easily computed from

$$\Delta P_{viscous}(\%) = 100 \left(\frac{1}{1+Fo}\right)$$

(12)

$$\Delta P_{inertial}(\%) = 100 \left(\frac{Fo}{1+Fo}\right)$$

(13)

As mentioned before, the range to classify the flow regime (laminar or turbulent) based on the Reynolds number is different from that used with basis on the Forchheimer number to determine the contribution of viscous and inertial effects. Laminarity is usually attained to $Re<1-10$ while the highly chaotic flow associated to turbulence emerges only for $Re>150-300$ (Seguin et al., 1998a, Seguin et al., 1998b and Hlushkou and Tallarek, 2006).

MATERIAL AND METHODS

Filtering Media

Four fiber-based filtering media of commercial grade and shaped as flat disks were tested in this work: polyester (PE) and polypropylene (PP) (*Gino Cacciari Indústria e Comércio de Filtros Ltda, Brazil*), cellulose (Cel) (*Ahlstrom Brasil Indústria e Comércio de Papéis Especiais Ltda, Brazil*) and a stainless steel cloth (Met) (*Apexfil Indústria e Comércio Ltda, Brazil*). The fiber diameter (df) and surface porosity (εs) of each medium were acquired by SEM pictures (*Philips XL 30 FEG*) and image analysis (*Image Pro-Plus 7.0*). Membrane thickness (L) was measured by a digital caliper, while the areal mass density or grammage (Gf) and fiber density (ρf) were obtained from suppliers and literature (Donovan, 1985, Purchas and Sutherland, 2002 and Hutten, 2007). The packing density of membranes (α) was calculated based on data of Gf, ρf and L (Donovan, 1985 and Hutten, 2007). The volumetric porosity (ε) was obtained from $\varepsilon=1-\alpha$.

Permeability Evaluation

Experimental evaluation of airflow permeability was carried out at room temperature ($\sim20-30$ °C) in the apparatus schematized in Fig. 2. Three specimens of each filtering medium were tested for reproducibility. The specimen under test was clamped between rubber gaskets within a cylindrical chamber that provided a circular flow area (A) of 44 cm^2, relative to a useful medium diameter of 7.5 cm. In order to avoid deflection of the fabric surface with increasing air velocities, the filtering medium was supported on a

10-mesh steel wire screen, which was assumed not to introduce appreciable pressure drop in the system. Dry air was forced to flow perpendicularly across the specimen in stationary regime according to seven preset levels of absolute pressures: 93, 193, 293, 393, 493, 593 and 693 kPa. For each test, the differential pressure through the specimen ($Pi-Po$) was measured by a digital pressure transducer (*Gulpress 200,Instrutemp*) and recorded as a function of volumetric airflow rate (Q), measured with a rotameter and converted to superficial velocity by $vs=Q/A$. The collected data set was treated according to the least-square method using a parabolic model of the type: $y=ax+bx^2$, in which y is $\Delta P/L$ (from Eq. (4)) and x is the fluid velocity at the filter exit (vso). The permeability parameters of Forchheimer's equation (3) were then calculated from the fitted constants a and b, respectively, by $k_1=\mu/a$ and $k_2=\rho/b$.

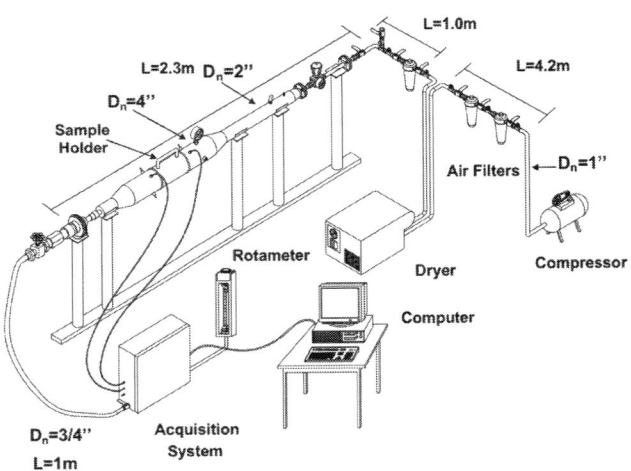

Figure 2: Scheme of permeation set-up used in the tests.

Both density (o) and viscosity (μo) of air were estimated with basis on the absolute pressure (Po) and temperature (To) at the filter outlet, respectively, by (Innocentini et al., 2005)

$$\rho_o = \frac{P_o MM_{air}}{Z_o RT_o}$$

(14)

$$\mu_0 = 1.73 \times 10^{-5} \left(\frac{T_0 + 273}{273}\right)^{1.5} \left(\frac{398}{T_0 + 398}\right)$$

$$(15)$$

in which *MMair* is the average molar mass of dry air (28.965×10^{-3} kg mol^{-1}) and R is the universal gas constant (8.314 Pa m^3 mol^{-1} K^{-1}). Z_0 is the compressibility factor based on the outlet filter pressure P_0. Experimental Z values retrieved from the literature for dry air at 298 K and pressures of 100–5000 kPa are in the range of 0.9999–0.9987 (Kadoya et al., 1985 and Perry and Green, 1999). Therefore, the ideal gas behavior ($Z=1$) was assumed valid within the pressure range experimentally tested in this work (0–693 kPa).

Simulation of Natural Gas Flow through Filters

Once the permeability parameters k_1 and k_2 were obtained from airflow experiments, simulations were performed to predict the pressure drop of the filter elements in operation with natural gas under absolute pressures (P_0) up to 5000 kPa and temperature of 25 °C. Molar composition of natural gas used for calculation of Z_0, o and μ_0 is shown in Table 1.

Table 1: Features of natural gas used in simulations

Component	Molar mass MMi(g/mol)	Mole fraction yi(%)	Critical temperature TCi(K)	Critical pressure PCi(kPa)
Methane (CH$_4$)	16.042	91.80	191	4599
Ethane (C$_2$H$_6$)	30.07	5.58	305	4872
Propane (C$_3$H$_8$)	44.1	0.97	370	4248
Butane (C$_4$H$_{10}$)	58.12	0.05	425	3796
Pentane (C$_5$H$_{12}$)	72.15	0.10	470	3370
Carbon dioxide (CO$_2$)	44.01	0.08	304	7383
Nitrogen (N$_2$)	28.13	1.42	126	3400

Apparent molar mass, *MMGN*=17.366 g/mol.

The compressibility factor Z_o, and viscosity (μo) of natural gas as a function of temperature and pressure were estimated from (Viswanathan, 2007 and Heidaryan et al., 2010)

$$Z_o = \ln\left(\frac{A_1 + A_3 \text{Ln}(P_{Pr}) + \frac{A_5}{T_{Pr}} + A_7(\text{Ln}(P_{Pr}))^2 + \frac{A_9}{T_{Pr}^2} + \frac{A_{11}}{T_{Pr}}\text{Ln}(P_{Pr})}{1 + A_2\text{Ln}(P_{Pr}) + \frac{A_4}{T_{Pr}} + A_6(\text{Ln}(P_{Pr}))^2 + \frac{A_8}{T_{Pr}^2} + \frac{A_{10}}{T_{Pr}}\text{Ln}(P_{Pr})}\right)$$

(16)

in which A_1–A_{11} are fitting coefficients given in Table 2, while PPr and TPr are the pseudoreduced pressure and pseudoreduced temperature, calculated, respectively, by (Heidaryan et al., 2010)

$$T_{Pr} = \frac{T_o}{\sum_{i=1}^{n} y_i T_{Ci}}$$

(17)

$$P_{Pr} = \frac{P_o}{\sum_{i=1}^{n} y_i P_{Ci}}$$

(18)

in which To and Po are the absolute pressure and temperature of the gas at the filter exit, respectively, yi is the mole fraction, TCi is the critical temperature and PCi is the critical pressure of each natural gas component i, as specified in Table 1.

Table 2: Fitting coefficients for Eq. (16) according to the range of PPr and TPr (Heidaryan et al., 2010)

Coefficient	0.2≤PPr≤3	3<PPr≤15
A_1	2.827793	3.252838
A_2	−0.468819	−0.130642
A_3	−1.262288	−0.644919
A_4	−1.536524	−1.518028
A_5	−4.535045	−5.391019
A_6	0.068951	−0.013796
A_7	0.190387	0.066006
A_8	0.620009	0.612078
A_9	1.838479	2.317431
A_{10}	0.405237	0.163222
A_{11}	1.073574	0.566059

The viscosity of natural gas was calculated as a function of temperature and pressure according to (Heidaryan et al., 2010)

$$\mu_o = 10^{-3}\ln\left(\frac{B_1 + B_2(MM_{NG}/T_o) + B_3(MM_{NG}/T_o)^2 + B_4\rho_o + B_5\rho_o^2 + B_6\rho_o^3}{1 + B_7(MM_{NG}/T_o) + B_8(MM_{NG}/T_o)^2 + B_9(MM_{NG}/T_o)^3 + B_{10}\rho_o}\right)$$

(19)

in which ρ_o is the gas density at P_o and T_o, MM_{NG} is the average molar mass of natural gas and $B_1–B_{10}$ are empirical constants given in Table 3. It worth mentioning that in Eq. (19) T_o is in °F, ρ_o in g cm^{-3} and MM_{NG} in g mol^{-1}, while μ_o is given in Pa s.

Table 3: Empirical constants for Eq. (19) (Heidaryan et al., 2010)

B_1	1.022872
B_2	−1.651432
B_3	5.757386
B_4	−0.073893
B_5	0.083891
B_6	0.297748
B_7	−1.451318
B_8	4.682506
B_9	1.918239
B_{10}	−0.098450

RESULTS AND DISCUSSION

Assessment of Physical and Permeation Features

The classification of fiber-based filtering media is usually attained to their packing density or solidity (α), defined as the fraction of the total volume of the filter element actually occupied by fibers. According toDonovan (1985), if $\alpha \leq 0.1$ (or $\varepsilon \geq 0.9$) the medium is classified as a fibrous filter, with a loose and open packing of unconnected fibers that offers small aerodynamic resistance to air flow and operates at face velocities in the 0.1–3 m/s range. On the other hand, for $\alpha \approx 0.1$–0.3 ($\varepsilon \approx 0.7$–0.9), the medium can be classified as a fabric filter, with a denser and planar packing of interlocked fibers that presents higher resistance to air flow and seldom exceeds a face velocity of 0.05–0.10 m/s.

The major properties and surface details of tested filters are given in Table 4 and Fig. 3. Packing density data of polymeric media shown in Table 4 are in the range 0.18–0.25, which classify them as fabric filters. The SEM micrographs shown in Fig. 3a–c confirmed the random two-dimensional fiber orientation typical of nonwoven felts PE, PP and Cel. Regions of melted fibers were also perceptible for the synthetic polymeric media (PE and PP). Typically used in industrial aerosol filtration (Donovan, 1985), these two media differ in fact mostly in their chemical resistance to strong alkali (higher for PP) and resistance to airflow temperature (higher for PE) (Hutten, 2007). Despite the similar fiber diameter and random fiber orientation, the cellulosic medium displayed smaller thickness compared to both PP and PE, a feature that results from the wet lay process by which the cellulose fiber web is made (Hutten, 2007, Donovan, 1985 and Purchas and Sutherland, 2002).

Table 4: Major features of tested filtering media

Filtering media	Polyester (PE)	Polypropylene (PP)	Cellulose (Cel)	Metallic (Met)
Texture	Felt	Felt	Felt	Woven
Areal mass density, Gf (g/m²)	550	600	150	535
Filter thickness, L (mm)	2.2	2.6	0.45	0.15
Fiber diameter, df (μm)	26	25	22	28
Fiber density, ρf (kg/m³)	1380	908	1500	7850
Surface porosity, εs (dimensioless)	0.80	0.76	0.74	0.30

Volumetric porosity, ε (dimensioless)	0.82	0.75	0.78	0.55
Packing density, α (dimensioless)	0.18	0.25	0.22	0.45

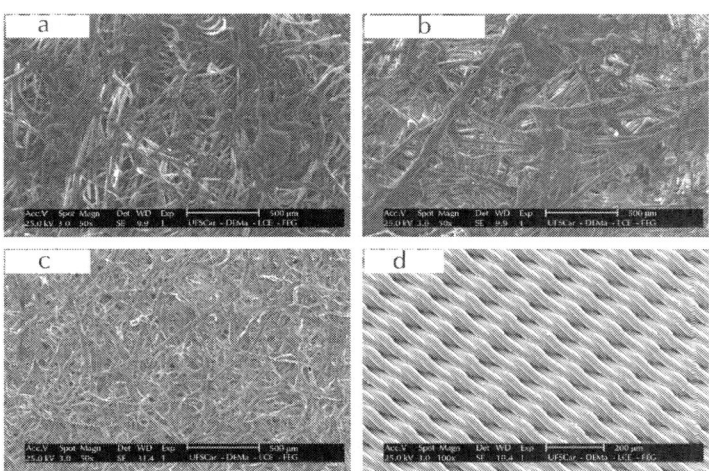

Figure 3: SEM details of tested filtering media.

The highest packing density (=0.45) and the smallest thickness (L=0.15 mm) of the metallic cloth given inTable 4 can be explained by its monolayered woven structure based on a dutch twilled weave pattern (Fig. 3d), characterized by the weft wires laid together as closely as possible, whereby one weft wire lies above and the next below the warp wire in each case.

Fig. 4 displays the normalized pressure drop curves (P/L) obtained from airflow tests with face velocities up to 0.21 m/s and absolute pressures up to 693 kPa. The fiber Reynolds number Ref was kept in the range 0.09–0.82 for all curves, characterizing the laminar flow. Nevertheless, the Forchheimer number Fo ranged from 0.09 to 1.43, which implied in a substantial and progressive contribution of the inertial term [vs^2/k_2] on total pressure drop (8–59%) with the increase of air velocity. This explains the clear parabolic trends of curves shown in Fig. 4 and allows one to conclude that Darcy's law is not valid as a predictive pressure drop model even for relatively low filtration velocities up to 0.2 m/s. Comparing the permeation curves for the four filtering media, the normalized pressure drop levels decreased in the following order: Met>Cel>PP>PE.

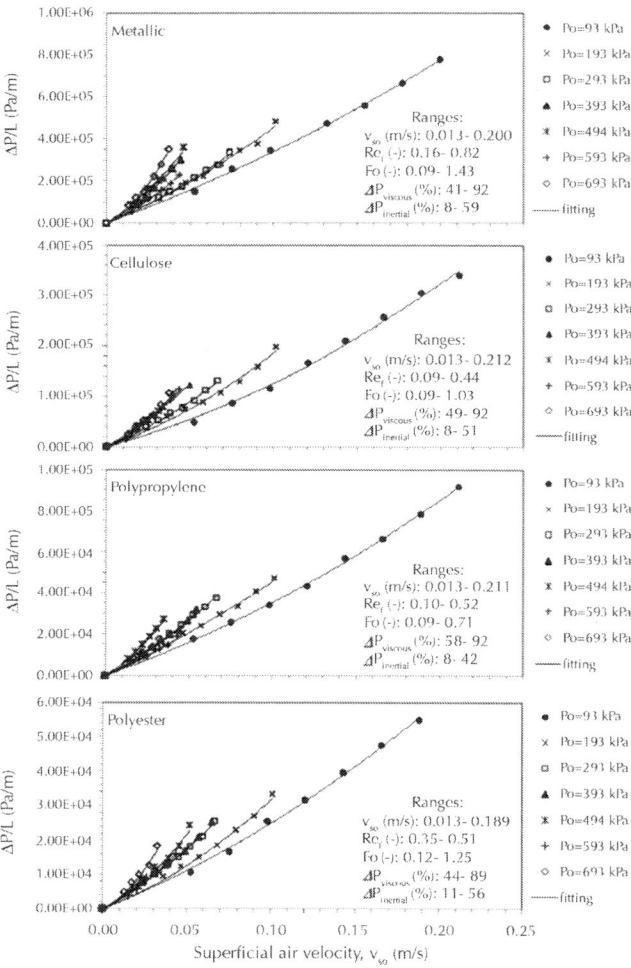

Figure 4: Experimental pressure drop curves for filtering media at different airflow pressures.

The increase in air pressure also shifted the pressure drop to higher levels, a behavior that could be related to changes in fluid (μ and) and medium (k_1 and k_2) properties. Air viscosity μ is not expected to vary significantly with pressure (<2% for the range 10–2500 kPa (Kadoya et al., 1985)). On the other hand, considering air as an ideal gas, its density varies linearly with pressure. Therefore, concerning the fluid features, the sevenfold increase in air pressure (93–693 kPa) did not affect the viscous energy dissipation represented by the linear term [$\mu vs/k_1$], but intensified proportionally the inertial losses given by the quadratic term [vs^2/k_2] of Forchheimer's equation. This also helps to explain the progressive warp of P/Lcurves with the increase in absolute pressure, as shown in Fig. 4.

The airflow conditions can also interfere with the permeable structure of flexible filtering media, with consequent changes in intrinsic permeability coefficients k_1 and k_2. Three potential types of flow-based media deformations are: (i) compression of individual filaments, (ii) compression of the fabric and (iii) deflection (stretching) of the fabric surface. Deformation of individual filaments is possible for polymeric hollow fibers in which air is trapped within the core cavities. The mechanical stress induced by non-isostatic pressure differences across the annulus may modify the cross-sectional fiber shape. Fabric deformation occurs for media exposed to high face velocities or to large differential pressures. If the fabric is supported by grids, pressure gradient along flow direction leads to an increase in packing density and therefore to a decrease in k_1 and k_2. If the fabric is unsupported, high flow rates leads to deflection and stretching of the fabric surface towards the flow direction, with increase of superficial porosity and of intrinsic permeability.

Fig. 5 shows that in fact none of the tested fiber media experienced physical deformations, since permeability coefficients remained practically unchanged up to 693 kPa. Such behavior seems reasonable, considering that no medium was tailored with hollow fibers, and suggests that the ranges of face velocity (0.01–0.21 m/s) and pressure drop (12–240 Pa) employed in the tests were not high enough to induce compression or deflection of the filtering fabrics. Average values for k_1 and k_2 increased in the following order: Met<Cel<PP<PE.

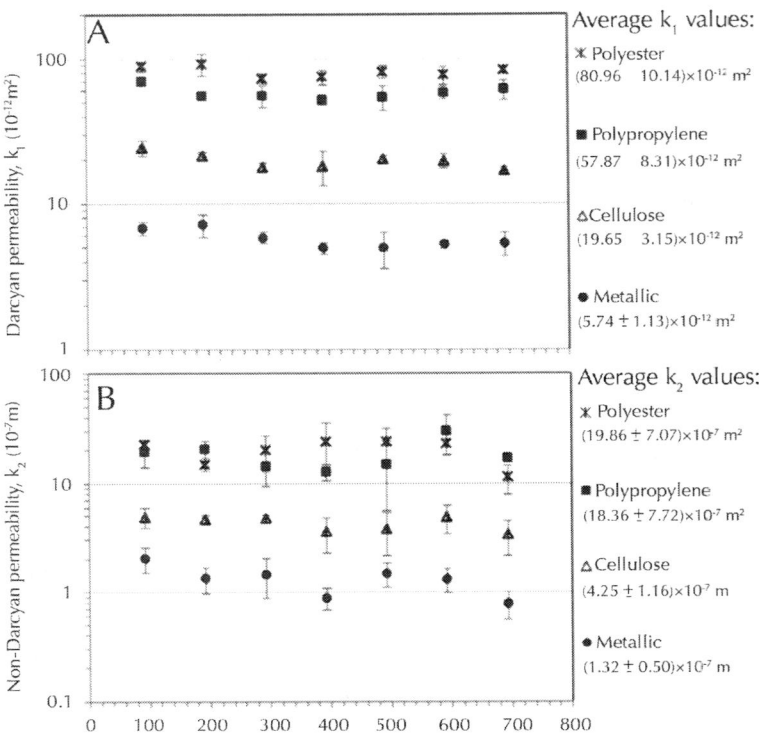

Figure 5: Darcian permeability (k_1); (B) non-Darcian permeability (k_2).

Fig. 6 locates these media in a comprehensive permeability map available in the literature. The ranges for k_1 (5.74–80.96×10^{-12} m^2) and k_2 (1.32–19.86×10^{-7} m) are typical of fiber media used in filtrations applications (Innocentini et al., 2005 and Hutten, 2007).

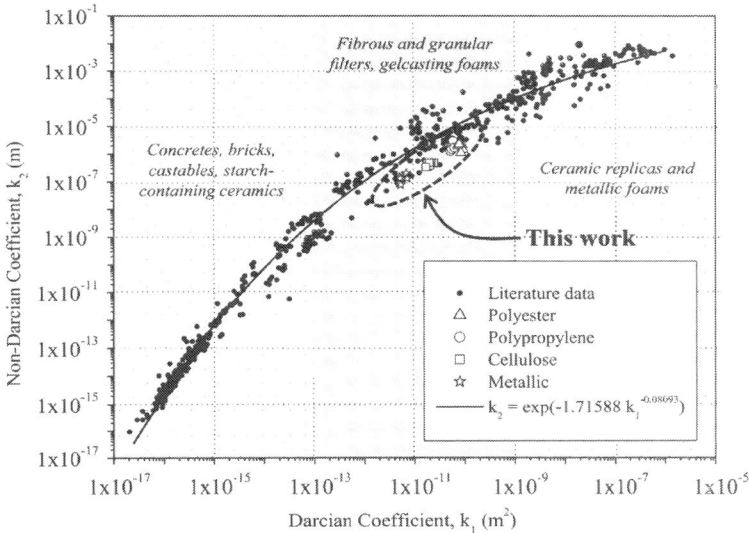

Figure 6: Location of experimental k_1 and k_2 data of filtering media in a comprehensive permeability map (Innocentini et al., 2005).

It was not intent of this study to perform a detailed comparison of experimental data with models available in the literature. However, it is instructive to distinguish the reasons for the different permeability levels of tested media as found in Fig. 5. Historically, both k_1 and k_2 have been empirically correlated to structural parameters of a variety of porous media by Kozeny–Carman or Ergun-like equations (Brown, 1993, Li, 1997, Brasquet and Le Cloirec, 2000, Innocentini et al., 2005, Wu et al., 2005, Wong et al., 2006, Wang et al., 2007 and Green et al., 2008). As a general rule, permeability increases with the increases in volume fraction of voids and in size of channels available for fluid flow. For permeability modeling of fiber-based media, the porosity (ε) has been often replaced by the packing density (α) and the channel (or pore) size replaced by other characteristic lengths, such as the aperture size (for screens and woven fabrics) and by fiber size (for nonwoven fabrics) (Li, 1997, Brasquet and Le Cloirec, 2000, Wu et al., 2005, Wong et al., 2006, Wang et al., 2007 and Green et al., 2008).

A classical equation to predict k_1 for fibrous pads valid in the range $0.06<\alpha<0.30$ (or $0.94>\varepsilon>0.70$) was developed by Davies (Hutten, 2007):

$$k_1 = \frac{d_f{}^2}{64\alpha^{1.5}(1+56\alpha^3)}$$

(20)

Davies development was based on flow past a single fiber of diameter df and on the hypothesis of laminar flow in which Darcy's law is valid (*Ref*<1 and *Fo*«1). The use of α and df data from Table 1 in Eq. (20)allowed reasonable predictions of k_1 for all tested media, as verified in Fig. 7. It included the metallic cloth, despite its woven structure and α beyond the validity range of Davies model. Therefore, considering that the filtering media displayed comparable fiber diameters (22 µm<df<28 µm), the differences in k_1 levels observed in Fig. 5 are fairly explained by the wider variation of packing density (0.18<α<0.45). The scatter can be related to clumps, fiber orientation and type of fiber (Hutten, 2007 and Donovan, 1985).

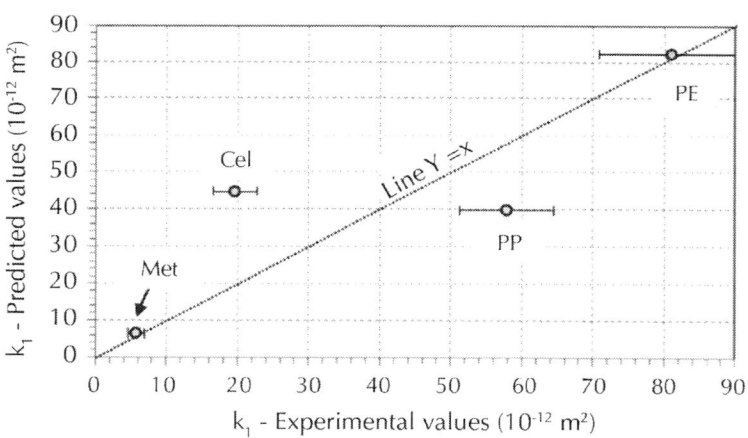

Figure 7: Comparison of experimental and predicted values for the Darcian permeability coefficients.

Another common form for expressing and comparing permeability of commercial textile fabrics, although fundamentally less rigorous, disregards the differences of medium thickness and specifies the resulting flow rate of air per unit area under a defined differential pressure across the specimen (Purchas and Sutherland, 2002). Metric version of this method proposed by the ASTM D 737 standard features permeability in units of $cm^3/s/cm^2$ at a minimum pressure drop of 125 Pa and 21 °C (ASTM, 1996). Such scale is helpful as it compares the air flow of commercial products based on the same pressure drop and on the actual thickness that each medium is sold and used.

Fig. 8 compares the air permeability for the four tested filtering media expressed according to ASTM D 737 (1996). Based on this scale, polyester and cellulose media displayed the highest permeability levels, followed by the metallic cloth and polypropylene. The apparent reduction of permeability with the increase of air pressure as seen in Fig. 8 is not related to changes in the fabric structure, as formerly discussed, but simply due to the increase in air density, which enhanced inertial effects on pressure drop.

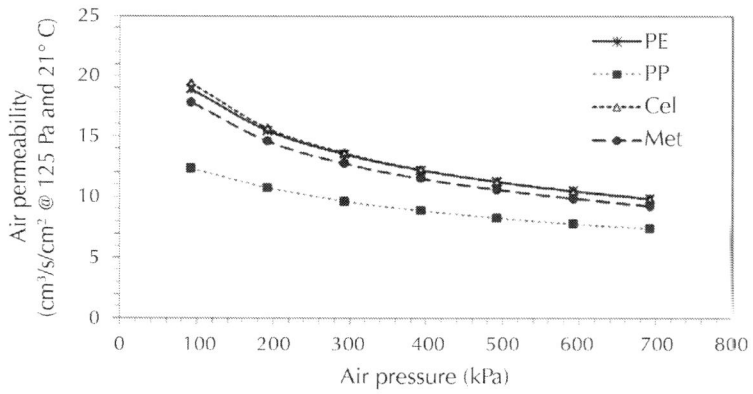

Figure 8: Permeability to air expressed according to ASTM D737 method as a function of air pressure.

Simulation of Natural Gas Permeation

The former section has established that the four tested filtering media were not structurally sensitive to gas pressure within certain limits of face velocity. Based on this achievement, a simulation of the high-pressure permeation behavior of natural gas through the four filtering media focused in this study was performed adopting similar limits. Properties of natural gas were based on the molar composition specified in Table 1and that is representative of the product available in the international market (Mokhatab et al., 2006). Table 5 compares the estimated properties μ and ρ of air and natural gas for room temperature (25 °C) and absolute pressures ranging from 100 to 5000 kPa. Air viscosity at high pressures was obtained directly from the literature (Kadoya et al., 1985) while air density was estimated from Eq. (14) with $Z=1$. On the other hand, the hypothesis of ideal gas was not assumed valid for natural gas and Eqs. (14), (15), (16),(17), (18) and (19) were used for evaluating the actual influence of pressure on density and viscosity.

Table 5: Properties of natural gas and dry air at 25 °C and different pressures

Pressure, Po (kPa)	Natural gas			Dry air		
	Zo	po	µo	Zo	po	µo
	(dimensionless)	(kg/m³)	(10⁻⁵ Pa s)	(dimensionless)	(kg/m³)	(10⁻⁵ Pa s)
100	1.00	0.70	1.11	1	1.17	1.85
500	0.99	3.54	1.12	1	5.84	1.85
1000	0.98	7.17	1.13	1	11.69	1.86
2500	0.95	18.50	1.16	1	29.22	1.88
5000	0.90	38.97	1.23	1	58.45	1.93

Two aspects are worth noting from data in Table 5: (i) viscosity of each fluid is little affected by pressure in the range 100–5000 kPa (11% increase for natural gas and 4% for air); (ii) for a same pressure, dry air presents both higher viscosity and density than natural gas. Since pressure drop through the filtering medium depends linearly on these both fluid properties, as stated by Eq. (3), two important conclusions are accordingly derived: (i) for a fixed face velocity, the increase of gas pressure does not affect significantly the viscous contribution $[\mu vs/k_1]$ of total pressure drop; (ii) for a same filtering medium and similar operational conditions, flow of natural gas will always result in lower pressure drops than flow of dry air. This latter statement is better visualized by simulated curves in Fig. 9, which compares the permeation behavior of the four filtering media with flow of either natural gas or dry air at a pressure of 5000 kPa and 25 °C. Curves were generated based on Eq. (3) fed with average permeability coefficients k_1 and k_2 retrieved from Fig. 5 and with fluid properties μ and ρ from Table 5. The limits of face velocity imposed in the simulations were similar to those reached in the actual experiments with airflow (0.013–0.212 m/s). Nevertheless, it is remarkable that the fiber Reynolds number (*Ref*) and the Forchheimer number (*Fo*) were both shifted to values far above those in which Darcy's law is considered valid, i.e., *Ref*<1 and *Fo*«1 (Innocentini et al., 2005). The approximate 50-fold increase in fluid densities at 5000 kPa, compared to values at ambient pressure (~100 kPa), induced equivalent increases in the contribution of the inertial-quadratic term $[\rho vs^2/k_2]$ on total pressure drop. Dry air is 50% denser than natural gas at 5000 kPa and for this reason produces proportionally higher pressure drop levels at the same face velocity.

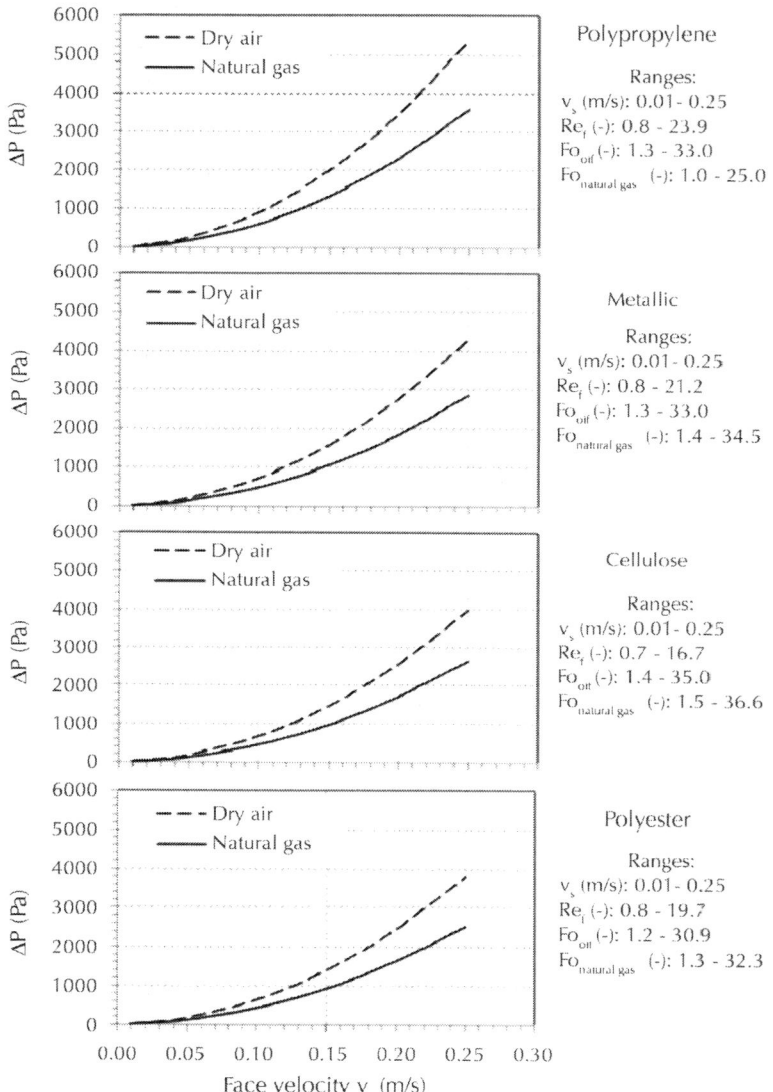

Figure 9: Simulation of pressure drop curves for dry air and natural gas at a flow pressure of 5000 kPa and T=25 °C.

Comparatively, permeation curves simulated for the four filtering media in Fig. 9 obeyed the previous trends observed in Fig. 8, with the resistance to flow decreasing in the following order:

PP>Met>Cel>PE. It is also instructive to compare this ordering with that observed in Fig. 4 (Met>Cel>PP>PE). Apparently contradictory, the divergent trends are in fact simply attained to the different ways by which permeability is expressed: the curves of pressure drop normalized by the medium thickness ($\Delta P/L$) shown in Fig. 4 are useful for the identification and comparison of filtering media that have different intrinsic permeability coefficients (k_1 and k_2). On the other hand, curves of pressure drop (ΔP) as shown in Fig. 9 give a direct measure of the net flow resistance that each medium produces when in operation, already including the contributions of its actual thickness and intrinsic permeability coefficients. Therefore, the filter with the highest permeability coefficients not necessarily yields the lowest pressure drop, as it also depends on the medium thickness. In the present study, the polypropylene fabric (PP) presented higher k_1 and k_2, but it was also much thicker than the metallic cloth (Met) and cellulose (Cel). The net effect was that, compared at the same face velocity, the PP fabric yielded higher pressure drop than Met and Cel media.

Recompression of natural gas in boosting stations is required to compensate the energy losses caused by friction in pipelines and also the losses due to the gas passage through filtering/scrubbing units and other non-pipe components. Because of the huge amount of gas transported worldwide, any procedure that provides the increase of filtration capacity without increasing the size of filtering elements or the pressure drop through them may result in relevant savings in power and operation costs (Wu, 1998 and Woldeyohannes and Majid, 2011).

The choice of ideal gas pressure for pipeline transportation and filtration is complex and involves aspects far beyond the scope of this work (Wu, 1998, Guo and Ghalambor, 2005, Kidnay and Parrish, 2006, Speight, 2007 and Wang, 2009). However, it is worth assessing the influence of gas pressure on the permeation performance of the four filtering media tested in this work. With this purpose, the pressure drop was simulated for radially inward flow through hollow cylinder elements, a typical geometry for natural gas filters. Predictions were based on Eq. (4) fed with average permeability coefficients k_1 and k_2 from Fig. 5and natural gas properties μ and from Table 5. The height H and outside diameter Do were fixed, respectively, in 0.250 m and 0.200 m, which resulted in an outside flow area $Ao=0.157$ m^2 common for all filters. The internal filter diameter (Di) was calculated by $Di=Do-L/2$ with basis on the actual thickness (L) of each medium given in Table 4. Gas velocity (vs) was kept fixed in the range 0.01–0.25 m/s, but absolute gas pressure was varied from 100 to 5000 kPa.

Some remarks are relevant from simulated curves in Fig. 10. The first is that gas becomes denser with the increase of pressure, thus a same face velocity (vs) results in progressively higher mass flow rates (w), as stated by Eq. (2). This increase of gas density also induces higher pressure losses due to inertial effects, as previously discussed. This explains why curves in Fig. 10 are shifted to higher ΔP and w values with the increase of gas pressure.

The economic benefits of operating natural gas filters at higher pressures are better demonstrated when curves in Fig. 10 are compared for a same pressure drop. For a given medium of fixed filtering area, higher permeation capacities may be achieved by increases of face velocity or increases of gas density. Disregarding the minor effect of pressure on gas viscosity, pressure drop is linearly dependent on ρ but in contrast it is proportional to the square of vs, especially for $Fo>1$. Thus, higher volumes of natural gas can be treated under a fixed pressure drop range if the filter unit operates at higher gas pressures and low face velocities rather than the opposite, i.e., low gas pressures and higher face velocities.

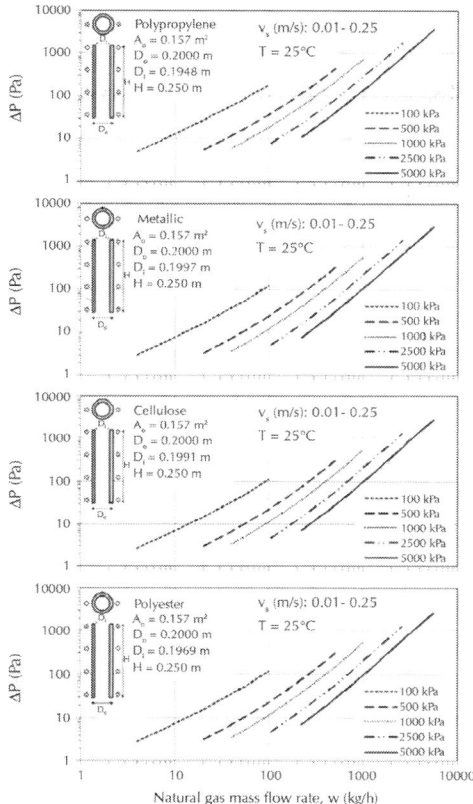

Figure 10: Simulation of pressure drop through cylindrical filtering elements for different gas pressures.

For instance, the polyester filter can produce a same pressure drop of P=100 Pa by operation at 100 kPa and vs=0.23 m/s or by operation at 5000 kPa and vs=0.05 m/s. In the former case, however, filter capacity is only 100 kg/h while in the latter it is increased to 1000 kg/h. Similar behavior is observed for the other three filtering media. Compared at the same pressure drop, the polyester medium yielded the highest permeation capacity and the polypropylene the lowest, though differences among all tested media were not so visually clear in Fig. 10 because of the logarithmic P scale.

CONCLUSIONS

Four filtering media commercially employed for removal of black powder in natural gas pipelines had their permeation behavior evaluated through high-pressure airflow experiments. Intrinsic permeability coefficients $k1$ and $k2$ remained practically unchanged for flow at pressures up to 693 kPa, indicating that no medium experienced physical deformations. The different permeability levels among the tested media were assigned to the differences in their packing density and fiber diameter. Despite the low velocities applied (0.01–0.21 m/s) and the fiber Reynolds numbers within the laminar regime (Ref.<1), the contribution of inertial effects on total pressure drop overcame 50%, as indicated by Forchheimer numbers above unity. Simulations compared the pressure drop levels for flow of dry air and natural gas in the range 100–5000 kPa. Dry air always resulted in higher pressure drops, as it is consistently denser and more viscous than natural gas in that pressure range. The four media were classified according to the following order of decreasing pressure drop: polypropylene > metallic cloth > cellulose > polyester.

Besides the specific achievements relative to the four tested filtering media, the analyses and simulations conducted in this work also allowed for some broader conclusions that are equally important and may serve as a guide for the development, selection and optimization of high-pressure filtering systems based on flexible media in the field of natural gas processing. Some of these conclusions are as follows:

- The compressibility factor Z of natural gas should not be disregarded from permeation modeling without detailed verification, as the influence of density (inertia) on pressure drop becomes progressively important with the increase of natural gas pressure.

- Air presents both higher viscosity and density than natural gas in any pressure range of importance for the gas industry, resulting therefore in a higher pressure loss under similar flow conditions. For this reason, any permeation data of filtering media acquired in laboratory scale should be presented in terms of intrinsic parameters, such as $k1$ and $k2$, which do not depend on the fluid properties and thus can be reliably used in pressure drop simulations in the actual flow conditions of interest.

- Within typical filtration velocities operated in industrial practice, pressure drop will be linearly dependent on natural gas density but proportional to the square of face velocity. Thus, higher volumes of natural gas can be treated under a fixed pressure drop range if the filter unit operates at higher gas pressures (higher gas densities) and lower face velocities.

- Darcy's law does not hold for typical high-pressure natural gas permeation, with deviations becoming more evident with the increase of gas density. A more appropriate model is Forchheimer's equation, which takes into account both viscous and inertial influences on pressure drop. The correct parameter to evaluate the validity range of Darcy's model is not the filtering velocity vs., but the Forchheimer number Fo.

- Flexible filtering media exposed to high pressure drops or based on hollow fibers may present structural deformations and affect permeation performance. An indirect and relatively easy laboratory procedure to verify such effect is the evaluation of the intrinsic permeability parameters $k1$ and $k2$ as a function of absolute gas pressure, such as the data presented in Fig. 5.

ACKNOWLEDGMENTS

Authors acknowledge the financial support from FAPESP and CNPq/ CT-Petro (Project no.: 16/2005). M.D.M. Innocentini also thanks UNAERP for supporting his research project

REFERENCES

1. ANEEL, 2008. Atlas de energia ele´trica do Brasil/Agencia Nacional de Energia ˆ Ele´trica. 3 ed. Brası´lia..

2. ASTM D737-96, 1996. Standard Test Method for Air Permeability of Textile Fabrics,

3. ASTM Designation: D 737-96. American Society for Testing and Materials, West Conshohocken, PA.

4. Azadi, M., Mohebbi, A., Scala, F., Soltaninejad, S., 2011. Experimental study of filtration system performance of natural gas in urban transmission and distribution network: a case study on the city of Kerman, Iran. Fuel 90 (3), 1166–1171.

5. Baldwin, R.M., 1998. Technical Assessment: Black Powder in the Gas Industry—Sources, Characteristics and Treatment. Report No. TA 97-4. Gas Machinery Research Council.

6. Brasquet, C., Le Cloirec, P., 2000. Pressure drop through textile fabrics— experimental data modeling using classical models and neural networks.

7. Chem. Eng. Sci. 55, 2767–2778. Brown, R.C., 1993. Air Filtration: An Integrated Approach to the Theory and Application of Fibrous Filters. Pergamon Press, Oxford.

8. Cornot-Gandolphe, S., Appert, O., Dickel, R., Chabrelie, M., Rojey, A., 2003. The challenges of further cost reductions for new supply options (pipeline, LNG, GTL). In: Proceedings of the 22nd World Gas Conference Tokyo, pp. 1–17.

9. DeShazo, P., Ladislaw, S., Primiani, T., 2007. Natural Gas, Energy Policy, and Regional Development: Brazil and the Southern Cone, Policy Papers on the Americas Volume XVIII. Study 1. Center for Strategic and International Studies.

10. Donovan, R.P., 1985. Fabric Filtration for Combustion Sources. Marcel Dekker, Inc., New York.

11. Green, S.I., Wang, Z., Waung, T., Vakil, A., 2008. Simulation of the flow through woven fabrics. Comput. Fluids 37, 1148–1156.

12.   Guo, B., Ghalambor, A., 2005. Natural Gas Engineering Handbook. Gulf Publishing Company, Houston, TX.

13.   Heidaryan, E., Moghadasi, J., Rahimi, M., 2010. New correlations to predict natural gas viscosity and compressibility factor. J. Pet. Sci. Eng. 73, 67–72.

14.   Hernandez-Rodriguez, M.A.L., Martinez-Delgado, D., Gonzalez, R., Unzueta, A.P., Mercado-Solis, R.D., Rodriguez, J., 2007. Corrosive wear failure analysis in a natural gas pipeline. Wear 263, 567–571.

15.   Hlushkou, D., Tallarek, U., 2006. Transition from creeping via viscous-inertial to turbulent flow in fixed beds. J. Chromatogr. A 1126, 70–85.

16.   Hutten, I., 2007. Handbook of Non-Woven Filter Media. Elsevier Science & Technology Books.

17.   IEA, 2010. Statistics—Natural Gas Information, 2010 edition International Energy Agency, Paris, France. M.D.M. Innocentini et al. / Chemical Engineering Science 74 (2012) 38–48 47

18.   Innocentini, M.D.M., Sepulveda, P., Ortega, F., 2005. Permeability. In: Scheffler, M., Colombo, P. (Eds.), Cellular Ceramics: Structure, Manufacturing, Properties and Applications. Wiley-VCH, Weinheim, pp. 313–340.

19.   Kadoya, K., Matsunaga, N., Nagashima, A., 1985. Viscosity and thermal conductivity of dry air in the gaseous phase. J. Phys. Chem. Ref. Data 14 (4), 947–970.

20.   Kidnay, A.J., Parrish, W.R., 2006. Fundamentals of Natural Gas Processing. CRC Press, Boca Raton, FL.

21.   Li, T., 1997. Dependence of filtration properties on stainless steel medium structure. Filtr. Sep., 265–273. (April).

22.   Mokhatab, S., Poe, W.A., Speight, J.G., 2006. Handbook of Natural Gas Transmission and Processing. Gulf Professional Publishing, Burlington, MA.

23.   Najibi, H., Rezaei, R., Javanmardi, J., Nasrifar, K., Moshfeghian, M., 2009. Economic evaluation of natural gas transportation

from Iran's South-Pars gas field to market. Appl. Therm. Eng. 29, 2009–2015.

24. Noman, R., Kalam, M.Z., 1990. Transition from laminar to non-Darcy flow of gases in porous media. In: Worthington, P.F. (Ed.), Advances in Core Evaluation: Accuracy and Precision in Reserves Estimation, pp. 447–462.

25. Perry, R., Green, D., 1999. Perry's Chemical Engineers' Handbook, 7th edition McGraw-Hill.

26. Purchas, D.B., Sutherland, K., 2002. Handbook of Filter Media. Elsevier Science & Technology Books.

27. Saremi, M., Kazemi, M., 2011. The effect of black powder composition on the erosion of compressor's impeller in gas transmission line. Adv. Mater. Res. 264–265, 1514–1518.

28. Scheidegger, A.E., 1974. The Physics of Flow through Porous Media, 3rd edition University of Toronto Press.

29. Seguin, D., Montillet, A., Comiti, J., 1998a. Experimental characterization of flow regimes in various porous media—I: limit of laminar flow regime. Chem. Eng. Sci. 53 (21), 3751–3761.

30. Seguin, D., Montillet, A., Comiti, J., Huet, F., 1998b. Experimental characterization of flow regimes in various porous media—II: transition to turbulent regime. Chem. Eng. Sci. 53 (22), 3897–3909.

31. Speight, J.G., 2007. Natural Gas: A Basic Handbook. Gulf Publishing Company, Houston, TX.

32. Tabkhi, F., 2007. Optimization of Gas Transmission Networks. Doctorate Thesis. Institut National Polytechnique de Tolouse.

33. Tanabe, E.H., Aguiar, M.L., Coury, J.R., 2011a. Performance of cellulose filter in gas filtration at high pressure conditions. In: Proceedings of the International Conference & Exhibition for Filtration and Separation Technology—FILTECH, Wiesbaden, Germany, II-94-II-101.

34. Tanabe, E.H., Barros, P.M., Bassan, K.R., Aguiar, M.L., 2011b. Experimental investigation of deposition and removal of

particles during gas filtration with various fabric filters. Sep. Purif. Technol. 80, 187–195.

35. Tanabe, E.H., Davoglio, G.R., Coury, J.R., Aguiar, M.L., 2010. Analysis of the polypropylene filter medium behavior at high pressure conditions to remove particulate matter in pipelines. In: Proceedings of the 19th International Congress of Chemical and Process Engineering CHISA, Prague, Czech Republic.

36. Trifilieff, O., Wines, T.H., 2009. Black powder removal from transmission pipelines: diagnostics and solutions. Presented at the Pipeline Rehabilitation & Maintenance Conference, Gulf International Convention Center, Bahrain, January

37. 19–21, 2009.

38. Tsochatzidis, N.A., Maroulis, K.E., 2007. Methods help remove black powder from gas pipelines. Oil Gas J. 105 (10), 1–9.

39. Victor, D.G., Jaffe, A.M., Hayes, M.H., 2006. Natural Gas and Geopolitics—From 1970 to 2040. Cambridge University Press, New York.

40. Viswanathan, A., 2007. Viscosities of Natural Gases at High Pressures and High Temperatures. M.Sc. Dissertation. Texas A&M University.

41. Wang, Q., 2007. An Investigation of Aerosol Filtration via Fibrous Filters. Ph.D. Thesis. North Carolina State University.

42. Wang, Q., Maze, B., Tafreshi, H.V., Pourdeyhimi, B., 2007. On the pressure drop modeling of monofilament-woven fabrics. Chem. Eng. Sci. 62, 4817–4821.

43. Wang, X., 2009. Advanced Natural Gas Engineering. Gulf Publishing Company, Houston, TX.

44. Weijermars, R., 2010. Value chain analysis of the natural gas industry—lessons from the US regulatory success and opportunities for Europe. J. Nat. Gas Sci. Eng. 2, 86–104.

45. Woldeyohannes, A.D., Majid, M.A.A., 2011. Simulation model for natural gas transmission pipeline network system.

Simulation Modelling Pract. Theory 19, 196–212.

46. Wong, C.C., Long, A.C., Sherburn, M., Robitaille, F., Harrison, P., Rudd, C.D., 2006. Comparisons of novel and efficient approaches for permeability prediction based on the fabric architecture. Compos.: Part A 37, 847–857.

47. Wu, S., 1998. Steady-state Simulation and Fuel Cost Minimization of Gas Pipeline Networks. Department of Mathematics. Ph.D. Thesis. University of Houston.

48. Wu, W.T., Liu, J.F., Li, W.J., Hsieh, W.H., 2005. Measurement and correlation of hydraulic resistance of flow through woven metal screens. Int. J. Heat Mass Transfer 48, 3008–3017.

49. Yamada, J., Kaneta, H., Nakayama, K., 2011. Analyses of Black Powder in Natural

50. Gas Pipeline. CORROSION 2011, March 13–17. Houston, Texas.

6

Energy Saving Opportunities in Integrated NGL/LNG Schemes Exploiting: Thermal-coupling Common-utilities and Process Knowledge

Mohd Shariq Khan[a], Yus Donald Chaniago[a], Mesfin Getu[b], and Moonyong Lee[a]

[a]School of Chemical Engineering, Yeungnam University, Dae-dong 712-749, Republic of Korea

[b]Chemical Engineering Department, Curtin University of Technology, CDT 250, 98009 Miri, Sarawak, Malaysia

ABSTRACT

Novel integrated and optimization schemes for natural gas (NG) liquid recovery and liquefaction step are presented. The liquefaction of NG was carried out using the newly developed KSMR liquefaction cycle while the separation of NG liquids was performed using energy efficient thermally coupled distillation schemes. The main highlight of integration are, (i) feed splitting to provide reboiler duty in the methane scrubber, (ii) common refrigeration utilities required all over the plant; and (iii) flexibility of integrated plant for easy transition between ethane recovery or ethane rejection. These integration steps minimize the overall plant-specific power requirements and the duplication of processing equipment. After successful integration, the MR cycle was optimized for the compression energy requirement by varying the refrigerant composition and operating pressure levels with the aid of an in-house established knowledge-based optimization (KBO) methodology. The KBO approach is robust in application and gives consistent results. Compared to the base case, 9% improvement in plant compression energy requirement was obtained.

INTRODUCTION

Clean burning characteristics and the ability to meet tough environment regulations has increased the demand for natural gas (NG) considerably, and a strong growth rate is expected over the next several years [1]. Primarily, NG is used as a fuel in the form of liquefied natural gas (LNG) and also serves as feed stock for the petrochemical industry [2]. Regardless of the method used to liquefy the NG stream, purification and removal of heavier hydrocarbons is needed prior to liquefaction. In the purification step, acid gases, water and other impurities are discarded. The heavier (C5+) recovery process has great economic benefit in terms of the valuable products obtained, such as propane, butane and other heavier hydrocarbon that are often sold separately.

In the LNG industry, several methods are available for NG liquefaction and natural gas liquids (NLG) recovery with a range of production capacities and product recovery, complexity, and operate independently in onshore platforms [3]. Meanwhile the rapid changes in the global NG market coupled with enthusiasm for exploiting offshore NG reserves, the emphasis in the LNG industry is now placed more on the integration of NGL and LNG for many of its benefit that it will bring to the overall plant economic performance [4]. The advancements and maturity in offshore technology for oil recovery paves the way for floating liquefied natural gas (FLNG) facility units. Although no such facility exists, the prospects for the FLNG industry have improved with the announcement of the Shell and Samsung joint venture for constructing multiple FLNGs [5]. The low space requirements, process safety, high efficiency, compactness, and simplicity of operation are a few of the main prerequisites of an offshore FLNG plant [6]. Considering the FLNGs requirements, it is important to develop a compact energy efficient integrated approach of the NG processing scheme for LNG and NGL recovery.

In this study, the concept of integrating NGL recovery with the NG liquefaction process was applied using Korea Single Mixed Refrigerant (KSMR) NG liquefaction as the base process [7]. This is a newly developed NG liquefaction cycle that was inspired by the operation of a single mixed refrigerant SMR cycle [8] and its attributes make it suitable for offshore plant application. Three representative NGL recovery sequences are integrated with this process and were analyzed for their energy efficiency, product recovery and purity. The base case of the NGL recovery sequence is the simple fractionation step, where three columns are used to separate methane/ethane, propane/butane and heavier hydrocarbons (C5+) as products. The remaining two sequences employed thermal coupling and a dividing wall column (DWC). The thermal coupling sequence is used because of its superior energy efficiency, and a DWC is employed because of its ability to allow reversible splits with no separation performed twice [7].

After successful integration, these integrated cases were analyzed for their energy efficiency improvements. Liquefaction is the most energy intensive step in the integrated facility. The liquefaction step provides the refrigeration duty for both the NG sub-cooling (LNG) and NGL recovery steps. Therefore, optimizing the liquefaction step will enhance the overall plant efficiency and profitability. The operating pressure levels of mixed refrigerant MR and its composition are the main decision variables in the KSMR cycle. The KSMR cycle is optimized with an in-house, established knowledge-based optimization (KBO) algorithm [9]. As the name indicates, the algorithm is inspired by the process knowledge and its characteristics, and it was used for optimization purpose. Using KBO methodology, the specific power required in the integrated schemes decrease significantly and saves considerable energy compared to the sub-optimally operated stand along facilities.

The aim of this paper was to propose several LNG and NGL integrated sequence and tailor the optimization scheme that may assist the personnel involved in offshore/onshore LNG projects with easy decision-making. The remainder of the paper is organized as follows. Section 2 provides an overview of a NG processing train, and Section 3 gives a brief review of NG liquefaction and recovery technologies. The KSMR cycle is described in detail in Section 3.2 and integration benefits are summarized in Section 4. Sections 4.1, 4.2 and 4.3 describe the proposed integrated schemes and Section 5 outlines the KBO algorithm. Main conclusions of the study are reported in Section 6.

NATURAL GAS PROCESSING TRAIN

Pretreatment Steps

Regardless of the methods used to separate NGL and liquefy NG and NGL separation, the purification of raw feed NG must be done before NG is sent to the cryogenic facility. The purpose of gas purification is to separate NG, condensate, non-condensable, acid gases and water to make these fluids suitable for sale or disposal [10]. Fig. 1 presents a schematic diagram of NG processing with the typical operating modules. In the first module, the physical separation of phases is performed such as gas, liquid water, liquid hydrocarbons, dirt or dust. The presence of two or three phases in the gas transmission lines causes liquid slugging, which in turn instigates mechanical problems, such as blanketing of the heat transfer surfaces and other process problems. Therefore, phase separators are installed before the remaining processing facilities. The inlet NG typically contains acid gases CO_2 and H_2S. The corrosive nature of these gases in the presence of water gave rise to an acidic aqueous solution and in addition, the CO_2 causes local freezing problem at low temperature process and hence must be kept below 50 ppm. These gases are removed to eliminate the undesirable components in the gas and satisfy the heating value of sales gases. The removal of acid gases can be performed using a reversible chemical reaction using physical absorption or employing fixed beds of solids [11]. After the successful removal of acid gases to the required limits, the feed NG is fed to the dehydration unit. Normally, the feed of NG to base load plants can contain water, as much as 1600 kg $H_2O/10^6$ m^3[12]. Under some conditions, NG can combine with water to form solid hydrates that can plug valves, fittings and pipelines, and decrease the heating value of NG [10]. Among the many drying methods, glycol absorption and molecular sieve dehydration are commonly used, and dehydration to dew points of 200 K can be accomplished using a molecular sieve. Once the NG is dried to the required value it is then sent to the mercury removal module. The presence of mercury results to aluminum corrosion and hence it should be removed from the feed gas. Normally, the mercury concentration, ranging from 1000 to 1800 ng/Nm^3, exists in most NG streams [12]. The removal of mercury is based on its high reactivity with sulfur and its compounds; and it is reduced to 10 ng/Nm^3.

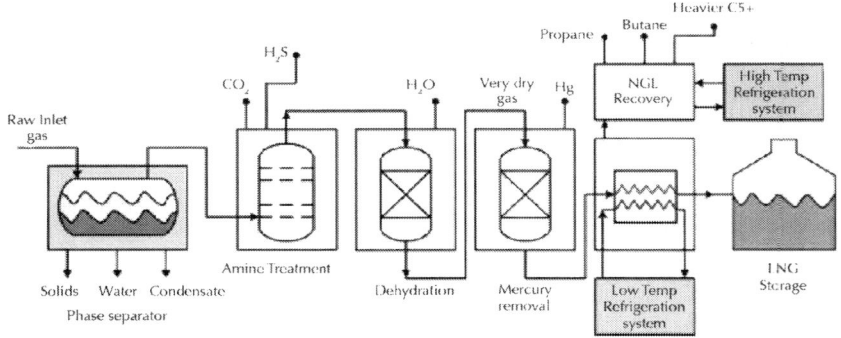

Figure 1: Block flow diagram for typical LNG plants.

NGL Recovery Step

NG received at the well head contains methane and a range of heavier hydrocarbons. Fractionation is required after the pretreatment step to recover those heavier HC liquids commonly referred as NGL (LPG; C3/C4 and condensate C5+). NGLs are removed to maintain the product dew points and yield a source of revenue. NGL has significantly greater value as separate products in their own right. The lighter NLG fractions, such as ethane, propane, and butane can be sold as a feedstock to refineries, whereas the heavier hydrocarbons can be used as gasoline-blending stock [10].

The heavier hydrocarbons are also removed for the following reasons:

- Pentane and heavier HCs with a high freezing point cause freezing and subsequent plugging of the process equipment during liquefaction.
- Heavy HCs (LPG and C5+) often have higher economic value than LNG because they serve as basic feed stock for petrochemical plants.
- Most LNG contracts specify a range of acceptable heating values. Therefore, required the removal of heavier HCs so that LNG does not exceed the upper limit on the heating value.

- Controlling the LNG heating value and selling heavier products in their own heating value range.

Natural Gas Liquefaction Step

When the feed NG is virtually free of NGL and contains predominantly methane, it enters the cryogenic LNG facility under elevated pressure. The gas is then subjected to several cooling stages by indirect heat exchange with the evaporating refrigerant until it is completely liquefied. The pressurized LNG is flashed to be stored at slightly above atmospheric pressure. Flashing generates boil-off vapor, which is utilized as a plant fuel or recycled within the process. NG liquefaction is an energy intensive process, and numerous methods of gas liquefaction have been developed [13] with the main differences being the type of refrigeration cycle used [14]. Similar to the NG liquefaction processes, a range of NGL recovery schemes have also been developed to accommodate a range of feed composition variations [15]. On the other hand, most schemes for NGL and LNG process development took place to work as standalone facilities, and not as a unified approach, which can include both NGL and LNG processing plants. To cover this information gap, several integrated schemes have been proposed and are described in Section 4.

Brief Review of Natural Gas Liquefaction and Recovery Technologies

Lim et al. reported a critically annotated review of the recent developments in NG liquefaction and optimization techniques [13]. The authors pointed out that very few studies have focused on performance improvements of the entire NG processing plant and their value chains. In another independent study by Getu et al. [16], potential process schemes available for NGL recovery were analyzed under different feed conditions. The economic assessment for each scheme was made by considering the capital, operating cost and profitability analysis. Some general conclusions regarding process performance under different conditions were made. The different processes discussed by Getu et al. were performed over a range of operating conditions by fixing common key process parameters and testing for various types of feed compositions that are classified as lean and rich. Under such circumstances, selecting the best process scheme could be a bit challenging since each process are designed under its operating conditions and this may be an advantage for that particular processes. Making valid comparisons between processes can be complicated because many processes are licensed or proprietary. Considering the above studies, very few studies have discussed the integration of NGL and LNG [17] because most of the work done has been proprietary [4]. The conclusions made from the above referred studies are that the integration can potentially produce both LNG and NGL products using significantly less energy than standalone facilities.

PROCESS MODELING AND HYPOTHESIS

The modeling of NG liquefaction (KSMR) and recovery (DWC, thermal coupling sequence) processes is performed in commercial process plant simulator Aspen Hysys™. Hysys extensive thermodynamic libraries and robustness in property calculations make this software well known in chemical engineering industry and academia. It has excellent steady-state process modeling capabilities and property prediction abilities for hydrocarbon system. The modeling of commercially available NG liquefaction [7], [13] and [14]and recoveries processes [15] and [16] are already performed using Hysys and its authority in designing novel LNG/NGL design configurations [18] as well as DWC [19] is also well established. Thus in the present study, Hysys steady-state modeling capabilities are exploited for the simulation of integrated process flow schemes. Note that to make process flow schemes more informative the Microsoft Visio was utilized for illustrating integrated schemes instead of using actual Hysys flowsheet. Hence not to be confused with the simple descriptive figures of integrated flow schemes because all the process schemes were rigorously modeled using Hysys simulation software.

Table 1 lists the simulation basis and feed conditions used in all cases. The approach temperature in the LNG exchanger was assumed to be 3°C [21]. Approximately 8% of LNG after flashing to atmospheric pressure was also assumed to be converted to boil-off-gas, which can be used as plant fuel or recycled back to the process. Furthermore, it was assumed that the fractionation columns were designed as the packed type and their heights were restricted to 20 equiv. trays because of their greater stability in a marine atmosphere. The desired product purities were constrained to 91% for LNG, 97% for LPG (C3, C4) and 99% for heavier (C5+). Note that the integrated approach presented in this paper was developed considering the offshore plant requirements. Nevertheless, when the space constraints are relaxed the design can accommodate the onshore plant requirements. When the restriction of the column height is no longer a constraint for onshore plant, then designing with the plant efficiency as a priority can be made.

Table 1: Feed condition and simulation basis

Property	Condition
NG feed condition	
Temperature (°C)	31.5
Pressure (bar)	63.7
Flow rate (kg/h)	344,720
NG feed composition	Mole fraction
Methane	0.9004
Ethane	0.0591
Propane	0.0201
i-Butane	0.0035
n-Butane	0.0057
i-Pentane	0.0018
n-Pentane	0.0017

Nitrogen	0.0033
Heat sink for refrigerant cooling	Sea water
Sink temperature (°C)	27
LNG production condition	
Temperature (°C)	−160.4
LNG tank pressure (bar)	1.06
Boil-off gas vapor fraction	8%
Compressor isentropic efficiency	0.75
Expander isentropic efficiency	0.75
Thermodynamic property calculations	
Vapor liquid equilibria	Peng–Robinson
Enthalpy calculations	Lee Kestler
Pressure drops	
LNG exchanger MR LK	1.0 bar
LNG exchanger MR HK	0.5 bar
Intercooler	0.5 bar

KSMR Liquefaction Cycle Suitability for Offshore Plant

In this paper, the concept of integrating NGL recovery with the NG liquefaction process is applied using Korea Single Mixed Refrigerant (KSMR) NG liquefaction as the base process. This is a newly developed NG liquefaction cycle, which is inspired by the simple operation of a single mixed refrigerant (SMR) process[14]. The SMR cycle uses a single loop of refrigerant mixture. Although the refrigeration duty can be satisfied with a portion of refrigerant expanded to very low pressure, all the refrigerant in the SMR process is expanded to the lowest working pressure to create a low temperature. Expanding all the refrigerants to the lowest working pressure leaves the SMR process energy intensive and limits its operation only to small scale NG fields. To overcome the small efficiency constraint of the SMR processes, several NG liquefaction cycles have been developed [13]. The Air Products & Chemicals Inc. propane precooled MR (C3MR) process is the most commonly used cycle for onshore operations [20]. In this cycle feed, NG is precooled with propane and liquefied, and sub-cooling is achieved through MR. On the other hand, the production capacity of this cycle is limited to 5 MTPA (million ton per annum) [7]. To remove the bottleneck in the C3MR process, the Dual Mixed Refrigerant (DMR) cycle was introduced, where precooling was performed using MR but its drawback included the large size and large amount of equipment. Considering the developments of the liquefaction cycles [13], most developments of the refrigeration cycles took place mainly for catering onshore operation and its needs. Offshore projects, however, have different requirements (high efficiency, size, safety, rapid startup and shutdown, etc.) and require a cycle tailored for its needs. Considering those needs, the KSMR cycle was developed and proposed as a simple, compact and power efficient cycle for the integrated operation of a NG processing plant in an offshore FLNG with a capacity of 2.5 Million Tonnes per annum (MTPA)[7].

KSMR Cycle Description

The main highlights of the KSMR cycle is that it uses different compositions of refrigerant streams generated from a single MR stream, at different pressure levels to minimize the compression energy requirement. Fig. 2 presents the conceptual KSMR cycle; a single MR stream is separated into heavy key (HK) and light key (LK) to act separately in a LNG exchanger. The HK precools the feed NG and warms the HK refrigerant, whereas LK precools warm LK and liquefies NG. The HK and LK streams exit the LNG exchanger at different pressures. They are re-compressed in separate compressor and restored to same mid pressure level. After mixing, the heavy components are separated and compressed separately to the mid-pressured state. The discharge pressures of LK/HK are the same but have different expansion pressures. The LK is expanded to the lowest possible pressure to provide sub-cooling, whereas the HK is expanded to relatively high pressures. This separation and expansion of refrigerants at different pressures results in significant compression energy savings. Further savings in the compression requirement in the process can be achieved by optimizing the inlet pressures. The appropriate composition of MR also affects the system optimality [22] and [23]. Hence, the MR compositions are also included in the optimization problem presented in Section 5.

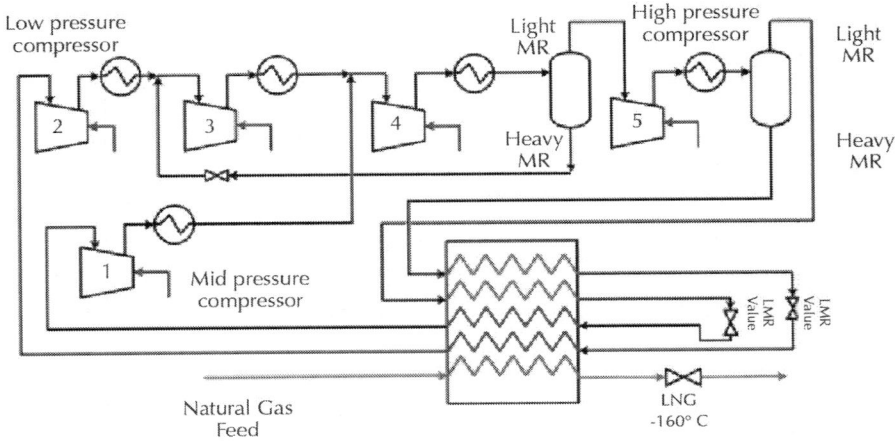

Figure 2: Process flow diagram KSMR cycle.

PROPOSED LNG AND NGL INTE-GRATION SCHEMES

To highlight the advantages of integrating a liquefaction section with a liquids recovery section, three cases of integration are presented in the remainder of the paper. It is noted that the simulation basis, feed conditions and other assumptions made for all the integrated schemes are commonly provided in Table 1. Moreover the rigorous modeling of all integrated scheme illustrated with Fig. 3, Fig. 4 and Fig. 5 are performed in commercial simulator Aspen Hysys (for details see Section 3). However simplified informative process flow schemes are used for better understanding and the rigorous model for any scheme can be retrieved using available information.

The material and energy balance for all the integrated schemes can be easily established using the stream properties (temperature, pressure, massflow depicted in the figures) and the information available in respective results Table 2, Table 3 and Table 4 of integrated schemes.

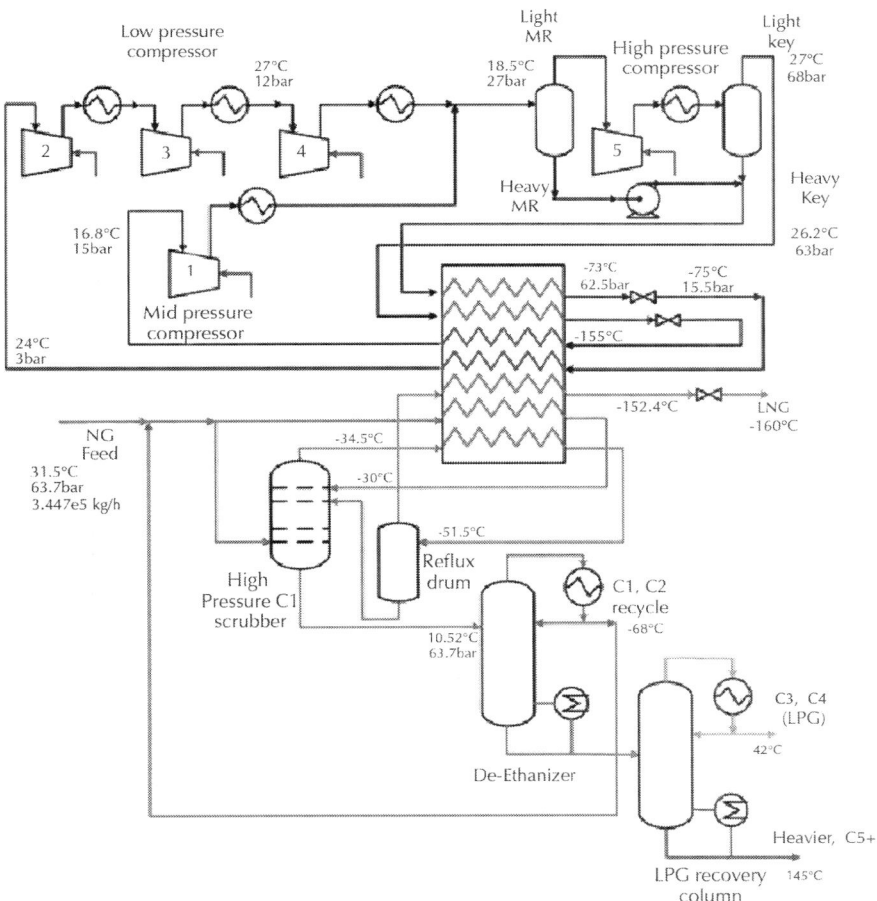

Figure 3: KSMR Integration with the conventional recovery sequence.

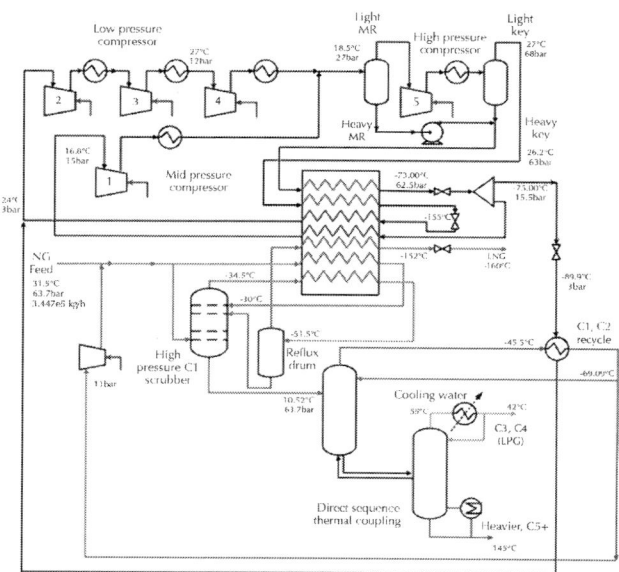

Figure 4: KSMR Integration with the thermally coupled direct sequence.

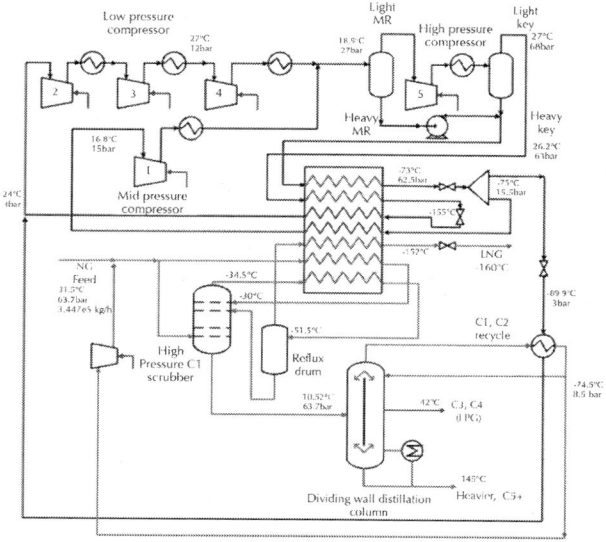

Figure 5: KSMR Integration with a dividing wall distillation column.

Table 2: Integration of the KSMR cycle with conventional separation sequence (base case)

Property	Value	Units	Remarks
NG feed flow rate	344,720.65	kg/h	
LNG flow rate	320,632.50	kg/h	
Condenser duty			
De-ethanizer	1870.01	kW	Refrigeration
LPG recovery column	2072.83	kW	Cooling water
Refrigeration duty provided by MR	0	kW	
Reboiler duty			
De-ethanizer	3093.01	kW	
LPG recovery column	2350.00	kW	
Components recovery			
Methane	0.9964	%	
Ethane	–		
Propane	0.9993	%	
i-Butane	0.9976	%	
n-Butane	0.9739	%	
i-Pentane	0.9071	%	
n-Pentane	0.9797	%	
Product purity			
LNG (C1,C2)	0.91	%	
LPG (C3, C4)	0.97	%	
Heavier (C5 +)	0.99	%	
Column hydraulics			
De-methanizer			
No of ideal trays	10		
Operating pressures	63.70 → 63.20	bar	
De-ethanizer			
No of ideal trays	10		
Operating pressures	11.20 → 11.00	bar	
LPG recovery column			
No of ideal trays	20		
Operating pressures	9.00 → 8.5	bar	

Table 3: Integration of the KSMR cycle with direct sequence thermal coupling for NGL separation

Property	Value	Units	Remark
NG feed flow rate	344,720.65	kg/h	
LNG flow rate	320,632.50	kg/h	
Condenser duty			
De-ethanizer	1418.76	kW	Refrigeration
LPG recovery column	1634.32	kW	Cooling water
Refrigeration duty provided by MR	1418.76	kW	
Reboiler duty			
De-ethanizer	–		
LPG recovery column	4547.57	kW	
Components recovery			
Methane	0.9932	%	
Ethane	–		
Propane	0.9989	%	
i-Butane	0.9946	%	
n-Butane	0.9743	%	
i-Pentane	0.9275	%	
n-Pentane	0.9796	%	
Product purity			
LNG (C1,C2)	0.91	%	Constrained
LPG (C3, C4)	0.97	%	Product purity
Heavier (C5 +)	0.99	%	
Column hydraulics			
De-methanizer			
No of ideal trays	10		
Operating pressures	63.70 → 63.20	bar	
De-ethanizer			
No of ideal trays	10		
Operating pressures	11.20 → 11.00	bar	
De-butanizer			
No of ideal trays	20		
Operating pressures	9.00 → 8.5	bar	

Table 4: Integration of KSMR cycle with a DWC column for NGL separation

Property	Value	Units
NG feed flow rate	344,720.65	kg/h
LNG flow rate	320,632.50	kg/h
Condenser duty		kW
DWC (de-butanizer + de-propanizer)	3369.59	
Refrigeration duty provided by MR	3369.59	kW
Reboiler duty		
DWC (de-butanizer + de-propanizer)	4861.31	kW
Components recovery		
Methane	0.9920	%
Ethane	–	
Propane	0.9999	%
i-Butane	0.9923	%
n-Butane	0.9651	%
i-Pentane	0.9103	%
n-Pentane	0.9827	%
Product purity		
LNG (C1,C2)	0.91	%
LPG (C3, C4)	0.97	%
Heavier (C5+)	0.99	%
Column hydraulics		
De-methanizer		
No of ideal trays	10	
Operating pressures	63.70 → 63.20	bar
DWC (de-butanizer + de-propanizer)		
No of ideal trays	20	
Operating pressures	9.00 → 8.5	bar

In all proposed schemes, feed splitting is applied in all cases, where a portion of the feed is directed to the high pressured methane scrubber (or wash tower) and the other portion is cooled in the LNG exchanger to be used as a reflux for the de-methanizer. Feed splitting eliminates the need for a reboiler in a de-methanizer and also decreases the feed precooling duty [7]. A reflux drum is also included in operation with a methane scrubber to minimize the losses of the heavier components in the top product. Therefore, the KSRM cycle with a methane scrubber is common in all proposed integrated sequences and the differences lay in the recovery of LPG (C3, C4+) and heavier (C5+) components, and will discussed in subsequent sections.

Few integration benefits

- The overall efficiency of the integrated process is higher than the individual standalone facilities.
- Plant throughput is increased with integrated schemes compared to standalone facilities.
- Heavier HCs have higher condensing temperatures than methane and will be liquefied in the liquefaction process.

KSMR Integration with Conventional NGL Recovery Sequence

Fig. 3 presents the first case of integration. This sequence is based on the KSMR cycle integration with conventional fractionation step, where a plurality of successive columns is used for components separation. As already mentioned in Section 4, a methane scrubber is integrated with a LNG exchanger to provide a cold reflux stream for heavier absorption. The cold and lean methane reflux enhances the NGL recovery efficiency within the column, and reduces the heavier components in the overhead stream to a minimum. The methane scrubber operates under high pressures to conserve the refrigeration compression requirements, but causes poor separation efficiency. Although this limitation cannot be avoided completely, the magnitude of excessive losses is certainly minimized using a common refrigeration utility in the de-methanizer. The condenser utility for the de-ethanizer and LPG recovery column is assumed to be generated independently. Therefore, KSMR integration with a conventional recovery sequence serves as the base case for a comparison with other integrated schemes and the comparison results are presented in Section 4.4. Table 2 lists the simulation results of KSMR integration with a conventional NGL recovery sequence. The product purities are constrained to 91%, 97% and 99% for LNG, LPG and heavier products in all the sequences studied to provide the same base for a fair comparison. The de-methanizer and de-ethanizer were designed to operate at 63.7 bar and 11.2 bar with 10 theoretical numbers of trays in each column, whereas the LPG column operates at 9 bar because this pressure allows the use of a high temperature utility in the condenser. As mentioned already, the column heights are constrained to 20 trays in each column, considering the operation of offshore plants.

The heavier hydrocarbons recovered from the methane scrubber are directed to the separation train, where the first column methane and ethane recovered at the top are recycled back in the process. The ethane is rejected in the LNG product but can be recovered easily based on its demand and price in the market. When the demand for ethane is high, increasing its recovery will generate additional revenue. On the other hand, ethane is often rejected in LNG when the market is depressed [13]. The bottom product of the de-ethanizer is sent to the LPG (C3, C4) recovery column where the top product consists of C3/C4, and C5+ are rejected in the bottom. Because the feed considered in the simulation basis is lean methane, the NGL products comprises only 7% of the entire NG feed.

KSMR Integration with Direct Sequence Thermal Coupling

Fig. 4 shows the KSMR integration with thermal coupling direct sequence (TCDS). The notable differences this sequence have compared to the conventional recovery integrated sequence (described in Section 4.1) are the use of energy efficient, thermally coupled columns for the separation of NGLs and the use of MR for providing condenser duties in a separation train. Table 3 lists the simulation results using this sequence. The production rate of LNG, NGL and column hydraulics are kept the same with the conventional sequence. Thermal coupling eliminates the use of one reboiler from the de-ethanizer but increases the reboiler load in the de-butanizer. The reflux condenser in the de-ethanizer column requires refrigeration, justifying the use of a mixed refrigerant from the cryogenic assembly. In the LPG column, the reflux condenser duty is satisfied with cooling water. Table 3 also lists the constrained product purities the required reboiler and condenser duties. The component recovery using this sequence are almost the same as those obtained in the previous sequence but the separation efficiency of i-pentane in the bottom is slightly higher than the previous sequence. The use of thermal coupling saves 24% of the condenser and 16% of the reboiler duties for about the same product purity compared to the conventional case.

KSMR Integration with a Dividing Wall Distillation Column

This sequence illustrated in Fig. 5 uses a DWC for the separation of three products, (i) C1/C2, (ii) LPG (C3/C4) and (iii) C5+ (i-pentane/n-pentane), from a single shell column. DWC is a fully thermally coupled column with a dividing wall installed in between, assisting in the sharp separation of three products by avoiding the re-mixing of the middle component. The simulation assumptions and results using this sequence are given in Table 4. The column operates at a pressure of 9 bar with 20 ideal trays that fall within the constrained limit (see Section 4). Similar to other conventional sequences, the methane scrubber is integrated within the LNG exchanger and the condenser duty in DWC is provided by the MR. The required constrained product purity (LNG 91%, LPG 97% and heavier 99%) is obtained with DWC but the elimination of a high temperature reflux condenser causes an increase in the duty of the low temperature refrigerant. This causes the over use of expensive cold utility thereby increasing the operating cost. To overcome incurred high operating cost, the thermally coupled columns are used proposed and described in detail in Section 4.2. Although thermally coupled columns results in saving of cold utilities however they need larger installation area. Thus depending on the project requirement, if space is at a premium, the DWC can be a better alternative than two separate columns coupled thermally.

NGL Recovery Sequences Comparison Results

Fig. 6 and Fig. 7 illustrate the NGL recovery sequence comparison results. Fig. 6 compares the reflux condenser duties for conventional and TCDS. Note that the reflux condenser duty for the LPG column inFig. 6 (for conventional and TCDS) is obviated, and comparison results present only the utilization of refrigeration and not water cooling (LPG column). TCDS results in a 24% decrease in the condenser duties compared to conventional sequence for the desired same purity. A comparison of the reflux condenser duties of all sequences including DWC showed an unusually high energy demand. This is because in a DWC sequence, a single reboiler and single condenser provides the entire boil-up and reflux condenser duty needed at the coldest and hottest temperature needed for separation. On the other hand, in the two column sequences (conventional and TCDS), the reboiler in the first and condenser in the second column can operate at intermediate temperatures. Therefore, the cold and hot utilities can be provided at an intermediate level that results in energy savings. When the reboiler duty is compared for all the sequences studied (see Fig. 7), the DWC can save approximately 10% reboiler duty, and a 16% in reboiler duty reduction was achieved using TCDS compared to the base conventional sequence. The use of DWC is justified only if the space is constrained due to the offshore project requirements. TCDS is thermally equivalent to a top-dividing wall column and can be replaced with a single shell operation like the DWC. Nevertheless, the problem of perfect insulation at the top of the column (required because of the large temperature difference between the different streams exist) is still an open challenge.

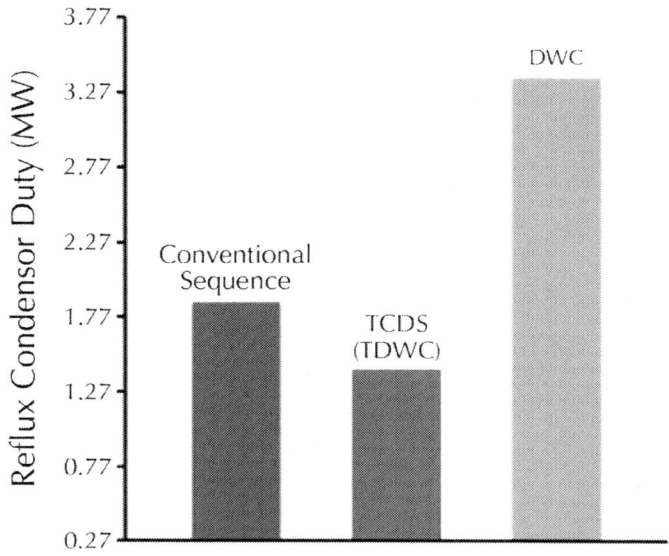

Figure 6: Reflux condenser duty comparison for all sequences (de-methanizer duty only for convention and TCDS).

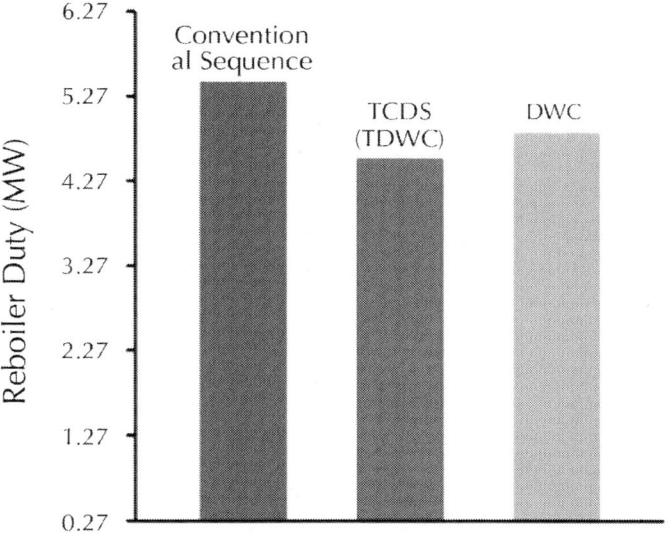

Figure 7: Reboiler duty comparison for all sequences.

PROCESS KNOWLEDGE-BASED OPTIMIZATION ALGORITHM OF LIQUEFACTION STEP

The liquefaction of NG is an energy intensive operation and accounts for as much as 30% of the entire energy use in NG processing plant [14]. This use of energy increases further in an integrated approach because the reflux condenser duties required for NGL recovery are also provided by the liquefaction plant. Thus sub-optimal operation of liquefaction plant results in substantial loss of utility cost. The optimal operation of the liquefaction plant strongly depends on the working refrigerant composition and its operating pressures and sub-optimal execution of these variables contribute to the process irreversibility. The main decision variables in the KSMR liquefaction plant are also the working refrigerant composition (nitrogen, methane, ethane, propane) and refrigerant operating pressures: mid pressure compressor (Comp-1) suction pressure, MR discharge pressure (Comp-5) and low pressure compressor (Comp-4) discharge pressure. The optimization of KSMR liquefaction plant was performed by considering compression energy minimization as objective. Knowledge of the boiling point difference between the MR components and their specific refrigeration effect was used to optimize the MR composition. The optimal composition corresponding to the minimum operating pressure in the MR cycle was searched with the aid of knowledge based optimization algorithm described in Fig. 5, details of knowledge-based optimization (KBO) methodology are reported in Khan et al. [9].

The main difference between KSMR liquefaction cycle and those reported in Khan et al. (SMR and C3MR)[21] are the operating pressure levels of working refrigerant. In the KSMR liquefaction cycle the working refrigerant is expanded to two different pressure levels. This difference of KSMR cycle in comparison to conventional cycles (SMR and C3MR) is made to expand only the required portion of refrigerant to the lowest possible pressure level (see Fig.

5, Comp-2 suction pressure, 3 bar). While the remaining refrigerant is expanded to the mid-pressure level (see Fig. 5, Comp-1 suction pressure, 15 bar). Thus, for optimizing the KSMR liquefaction cycle an additional step in algorithm reported in Khan et al. [21] is included to compensate for the structural difference in liquefaction cycles. In this additional step, the modified algorithm illustrated in Fig. 8 tries to find the maximum value of mid-pressure operating compressor in agreement with the rest of it (Comp-1 suction pressure, Fig. 5) within the feasible domain that helps in restoring refrigerant to high pressure discharge state (Comp-5 discharge pressure) that results in less compression energy. The reported process knowledge algorithm [21] was applicable for general MR system but its customized form is used in this study illustrated in Fig. 8. The compensation of structural changes in the KSMR liquefaction cycles results in the customized forms (Fig. 8). However the basic form with refrigerant composition management remains the same.

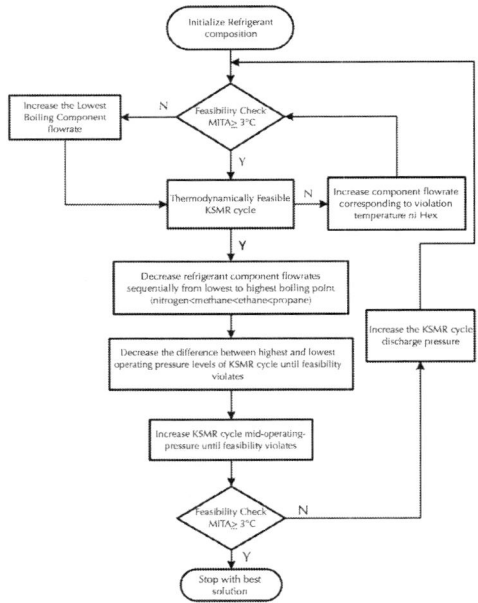

Figure 8: Knowledge-based optimization algorithm for the KSMR liquefaction cycle.

Optimization Results of Liquefaction Step

The KBO methodology unlike other optimization techniques is free from the good initial estimates and can commence from any point. To illustrate the optimization results a base case is selected, and the comparison results are presented in Table 5. The optimization procedure was applied to the KSMR/TCDS integrated sequence and the compression energy minimization was selected as an optimization objective. Optimization begins by manipulating the MR composition using algorithm described in Fig. 8. The optimization results revealed a 9% decrease in the compression energy requirement and also specific energy requirement for NG liquefaction in the KSMR/TCDS integration sequence compared with the base case. A comparison with Vatani et al. [18] showed that the proposed integrated plant required 6.6% less specific energy (Table 6). A higher overall operating pressure of the MR cycle is predicted because of the high compression energy demand in the base case. In the optimized case, the suction pressure of Comp-1 increases and Comp-5 discharge pressure decreases, resulting in an overall decrease in the system operating pressure that decreases the system irreversibility and compression energy. The decrease in the high boiling ethane and propane flow rate compensate for the narrow gap (low irreversibility) between the composite curves presented in Fig. 9 and Fig. 10.

Table 5: Optimization of the KSMR liquefaction cycle using the KBO algorithm

Property	Base case	Optimized case
Compression duty (kW)	135,969.82	123,864.90
Specific power requirement (kW/kg-LNG)	0.4241	0.3863
Approach temp (°C)	3.15	3.00
	Values	
Compressor-1 suction pressure (bar)	9	15
Compressor-4 discharge pressure (bar)	29	27
Compressor-5 discharge pressure (bar)	71	68

Nitrogen flow rate (kg/h)	105,000	89,200
Methane flow rate (kg/h)	210,000	225,600
Ethane flow rate (kg/h)	472,000	401,900
Propane flow rate (kg/h)	699,000	607,500

The bold and italic value of Specific power requirement are the optimization objective. The optimized case represents small specific power demand compared to the base case.

Table 6: Specific energy comparison

Property	Vatani et al. [18]	Proposed KSMR/TCDS
Specific power requirement (kW/ kg-LNG)	0.4140	0.3863

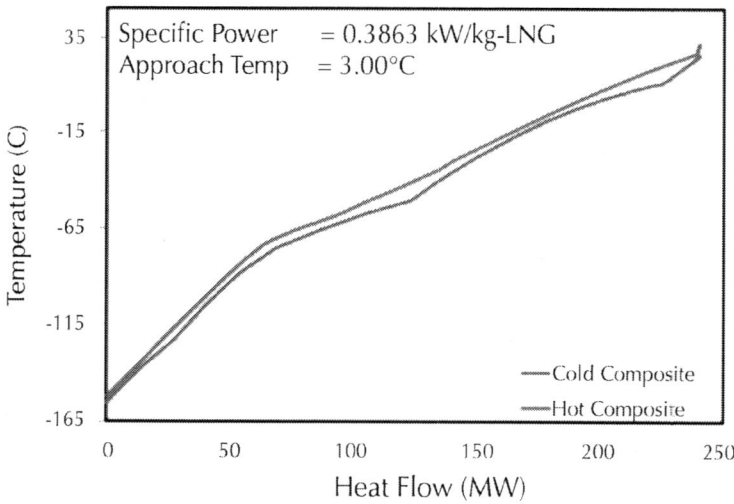

Figure 9: Composite curves in a LNG exchanger for KSMR/TCDS integration.

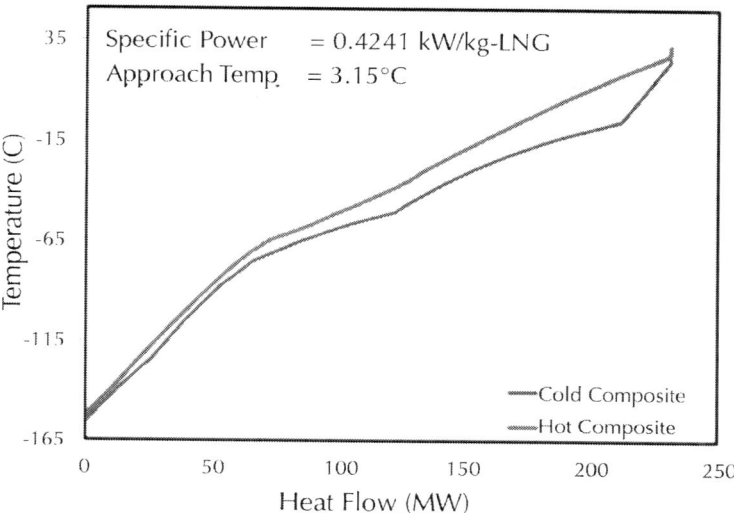

Figure 10: Composite curves in a LNG exchanger for conventional integration sequence.

CONCLUSIONS

The integration scenarios of a newly developed KSMR NG liquefaction cycle with three NGL recovery sequences are presented. Careful integration of the LNG and NGL recovery plant successfully eliminates the separate refrigeration utility requirement in NLG fractionation and results in a lower overall specific energy requirement in an integrated plant. The conventional NGL recovery sequence results in deep NGL recovery at the expense of higher reboiler and condenser duties. The required product purities can be achieved using the thermally coupled sequence for NGL recovery. Therefore, they are proposed for integration with the KSMR plant. The simulation results predict that among the three studied sequences, the TCDS or top dividing wall column performs better in terms of the reboiler and condenser duty and increases the net present value of the NG processing plant. The use of DWC is justified only if the space is constrained due to offshore project requirements. TCDS is thermally equivalent to the top-dividing wall column and can be replaced with a single shell operation like DWC. Nevertheless, the problem of perfect insulation at the top of the TCDS still remains a challenge so the use of DWC is still open for offshore platforms.

Finally, the in-house established MR system knowledge-based optimization approach was employed to minimize the compression energy requirements in the integrated plant. Compared to the base case, a 9% decrease in compression energy requirement is obtained. Using KBO, it was found that oversetting of the discharge pressure of MR and the high flow rate of higher boiling components in MR contributes mostly to the system irreversibility and causes high compression requirements. The use of KBO algorithm found to be robust in application and brings consistent results. The proposed integrated sequences can satisfy the major offshore LNG project requirements and paves the way for the successful exploitation of offshore NG fields.

ACKNOWLEDGMENTS

This study was supported by a grant from the Gas Plant R&D Center funded by the Ministry of Land, Transportation and Maritime Affairs (MLTM) of the Korean government, and by Basic Science Research Programthrough the National Research Foundation of Korea (NRF) funded by the Ministry of Education, Science and Technology (2012012532).

REFERENCES

1.   International Gas Union, World LNG Report, 2013.

2.   A.J. Kinday, W.R. Parrsih, Fundamentals of Natural Gas Processing, Taylor and Francis, Boca Raton, 2006.

3.   S. Kumar, H.T. Kwon, K.H. Choi, W. Lim, J.H. Cho, K. Tak, LNG: an eco-friendly cryogenic fuelfor sustainable development,Appl. Energy 88 (2011) 4264–4273.

4. H.M.Hudson,J.D.Wilkinson,K.T.Cuelar,M.C. Pierce,Integratedliquids recovery technology improves LNG production efficiency, in: 82nd Annual GPA Convention, Texas, March, 2003.

5. Press Release, Technip, Shell and The Technip Samsung Consortium Sign Agreement to Strengthen Floating LNG collaboration, December 2012.

6. M.A. Barclay, C.C. Yang, Offshore LNG: the perfect starting point for the 2-phase expander? in: Offshore Technology Conference, May, 2006, p. OTC 18012.

7. S. Lee, N.D.V. Long, M.Y. Lee, Design and optimization of natural gas liquefaction and recovery processes for offshore floating liquefied natural gas plants Ind. Eng. Chem. Res. 51 (2012) 10021–10030.

8. L.K. Swenson, Single Mixed Refrigerant Closed Loop Process for Liquefying Natural Gas, 1977. US Patent 4,033,735.

9. M.S. Khan, S. Lee, G.P. Rangaiah, M.Y. Lee, Knowledge based decision making method for the selection of mixed refrigerant systems for energy efficient LNG processes, Appl. Energy 111 (2013) 1018–1031.

10. S. Mokhatab, W.A. Poe, J.G. Speight, Handbook of Natural Gas Transmission and Processing, Gulf Publication, Burlington USA, 2006.

11. J.G. Speight, B. Ozum, Petroleum Refining Processes, 1st ed., Taylor & Francis, Basel, Switzerland, 2009.

12. K. Othmer, Encyclopedia of Chemical Technology, 5th ed., 2004.

13. W. Lim, K. Choi, I. Moon, Current status and perspectives of liquefied natural gas (LNG) plant design, Ind. Eng. Chem. Res. 52 (2013) 3065–3088.

14. M.S. Khan, S. Lee, M.Y. Lee, Optimization of single mixed refrigerant natural gas 827 liquefaction plant with nonlinear programming, Asia-Pac. J. Chem. Eng. 828 (7 (Suppl. 1)) (2012) S62–S70.

15. M. Getu, M.S. Khan, N.D.V. Long, M.Y. Lee, Studying the effect of feed composition variation on typical natural gas liquid (NGL) recovery processes, Comp. Aided Chem. Eng. 31 (2012) 405–409.

16. M. Getu, S. Mahadzir, N.V.D. Long, M.Y. Lee, Techno-economic analysis of potential natural gas liquid (NGL) recovery processes under variations of feed compositions, Chem. Eng. Res. Des. 91 (2013) 1272–1283.

17. D.G. Elliot, Benefit of integrating NGL extraction and LNG liquefaction technology, in: 5th AIChE National Spring Meeting, 2005.

18. A. Vatani, M. Mehrpooya, B. Tirandazi, A novel process configuration for coproduction of NGL and LNG with low energy requirement, Chem. Eng. Process. Process Intensif. 63 (2013) 16–24.

19. R. Premkumar, G.P. Rangaiah, Retrofitting conventional column systems to dividing-Wall Columns, Chem. Eng. Res. Des. 87 (2009) 47–60.

20. J.L.S. Gaumer, C.L. Newton, Combined Cascade and Multicomponent Refrigeration System and Method, 1973. US Pat. 3,763,658.

21. M.M. Hasan, I.A. Karimi, H.E. Alfadala, H. Grootjans, Operational modeling of multistream heat exchangers with phase changes, AIChE 55 (2009) 150–171.

22. M.S. Khan, M.Y. Lee, Design optimization of single mixed refrigerant natural gas liquefaction process using the particle swarm paradigm with nonlinear constraints, Energy 49 (2013) 146–155.

23. M.S. Khan, S. Lee, M. Hasan, M.Y. Lee, Process knowledge based opportunistic optimization of the N2-CO2 expander cycle for the economic development of stranded offshore fields, Journal of Natural Gas Science and Engineering 18 (2014) 263–273.

7

Swirling Flow of Natural Gas in Supersonic Separators

Chuang Wen, Xuewen Cao, and Yan Yang

Department of Oil and Gas Engineering, China University of Petroleum, Qingdao 266555, China

ABSTRACT

The supersonic swirling separator is a new apparatus for offshore and subsea natural gas separation, due to its lightweight and the viability of unmanned operation. A new supersonic swirling separator was designed for the numerical calculation, in which a central body was inserted based on the principle of conservation of angular momentum. Axial and radial distribution of the main parameters of natural gas flow was investigated with *RNG K-ε* turbulence model.

The effects of the shock waves on the natural gas flow fields were analyzed in the supersonic separator. The results show that water and heavy hydrocarbons can be separated from natural gas due to the low temperature and high centrifugal field. The non-uniformity of radial distribution of the gas dynamic parameters significantly affects the gas/liquid separation. The position of the shock wave determines the distribution of the temperature, which has a great influence on the re-evaporation of liquid droplets.

INTRODUCTION

Natural gas is usually saturated with significant quantities of water, which must be removed for gas transmission. The natural gas industry has recognized that dehydration is necessary to ensure smooth operation of gas transmission lines. Dehydration prevents the formation of gas hydrates and reduces corrosion. Unless gases are dehydrated, liquid water may condense in pipelines and accumulate at low points along the line, reducing its flow capacity [1]. Therefore, several methods have been developed to dehydrate gases, such as cooling, adsorption and absorption. However, the traditional natural gas processing has proven both costly and complex, requiring large facilities with high operating expenses, and even could cause environment problem due to the injection of hydrate inhibitor.

The supersonic swirling separator has been introduced to treat natural gas for condensing and separating water and heavy hydrocarbons [2], [3], [4] and [5]. It is a compact tubular separator device with no moving parts, enabling high reliability and availability. It is smaller, lighter, cheaper and with fewer emissions than conventional dehydration plant [6], [7] and [8]. It is suited for platforms, due to its lightweight and the viability of unmanned operation. Significant potential has been identified for future application of this technology on various other gas processing separation applications including deep LPG extraction, bulk removal of CO_2 and H_2S, and subsea gas processing. A numerical simulation was performed for the water vapor condensable supersonic flows through Laval nozzles under different flow friction conditions [9]. The mixing flow field of the nitrogen and water vapor is numerical simulated in an axial flow supersonic nozzle [10]. The natural gas behavior is illustrated when it is considered to be real and how erroneous the properties may become when the gas is assumed to be ideal in axial flow supersonic nozzles [11]. Effect of nozzle geometry is discussed by inserting a constant area channel between the convergent and divergent parts of the axial flow nozzle [12]. Design methods of axial flow supersonic nozzles for natural gas separation are numerical calculated, and the flow field is obtained [13].

But there are few reports about the radial distribution of gas dynamic parameters and shock waves in supersonic separators. Jassim et al. [11] investigated the effects of the gas model on the shockwave positions of two different gases (methane and nitrogen). The results show that the shock position of nitrogen is in front of methane, when the ideal gas model is used. However, shockwave occurs earlier for methane with the real gas model. Karimi and Abdi [14] studied the influences of the inlet pressure, temperature and back pressure on the shock positions. It indicates that shock occurs earlier with the increases of the inlet or back pressure. The high inlet temperature delays the position of the shockwave.

The purpose of this paper is to investigate the effects of the supersonic swirling flow on the radial distribution of the main parameters of gas flow and the effects of shock position.

NEW GEOMETRY OF THE SEPARATOR

In this paper, a new supersonic swirling separator is designed which incorporates a central body, allowing the principle of conservation of angular momentum to be harnessed. The channel between the wall and the central body forms a annular nozzle, a cyclonic separation section and a diffuser, as seen in Fig. 1. The swirling motion is generated by the vanes, which turned through 45° and located in the subsonic part of the channel on the central body axis.

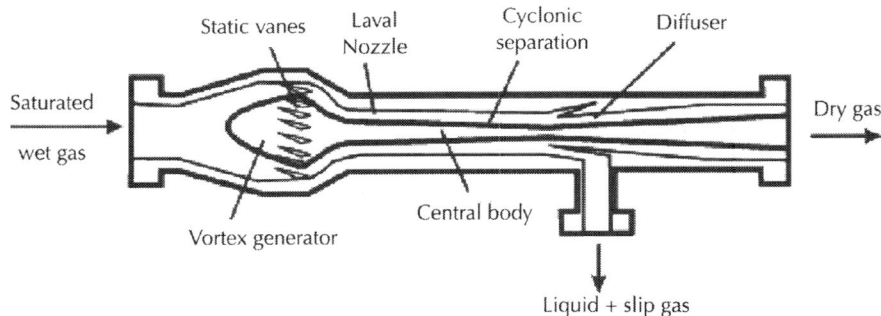

Figure 1: Schematic diagram of a supersonic swirling separator.

The channel between the wall and the central body forms a new annular convergent–divergent nozzle, which is composed of three sections: the convergent (subsonic zone), throat area (critical zone) and divergent section (supersonic zone). In the convergent section, the flow is accelerated and the sonic velocity is reached at the throat. In this research, the dimensions of the convergent section of the nozzle are calculated by Eq. (1).

$$\begin{cases} \dfrac{D - D_{cr}}{D_1 - D_{cr}} = 1 - \dfrac{1}{X_m^2}\left(\dfrac{x}{L}\right)^3 \quad \left(\dfrac{x}{L} \le X_m\right) \\[4mm] \dfrac{D - D_{cr}}{D_1 - D_{cr}} = \dfrac{1}{(1 - X_m)^2}\left(1 - \dfrac{x}{L}\right)^3 \quad \left(\dfrac{x}{L} > X_m\right) \end{cases}.$$

$$(1)$$

where D_1, Dcr and L are the inlet diameter, the throat diameter and the convergent length, respectively. Xm is the relative coordinate of this convergent curve. Typically, $Xm = 0.3 - 0.7L$. x is the distance between arbitrary cross section and the inlet, and D is the convergent diameter at arbitrary cross section of x.

The divergent section of the nozzle can further accelerate the flow to supersonic velocity with resulting in lower temperatures, which leads to the nucleation of water and hydrocarbons. However, under certain conditions, a shock wave can occur, destroying the low temperature. Therefore, it is necessary to prevent the formation of the shock waves in the divergent channel of the nozzle for this apparatus. The divergent angle of the nozzle is specified to be 4° in the numerical calculation.

When the gas liquid mixture passes through the cyclonic separation section, the liquids goes into the drainage pipe. The gas finally moves across the diffuser within which the remaining kinetic energy in the separated flows is transformed to increase the static pressure. The divergence angle should be in an appropriate range because the shock can interact with the boundary layer, and this can delay the transition from supersonic flow to subsonic flow even further. A conical diffuser is used to recover the pressure by decelerating the axial component of the velocity for this new design.

In our new design, the supersonic separator is about 587 mm long and 80 mm in the inlet diameter and 40 mm in the outlet diameter. The diameters of the central body and wall at the nozzle outlet are 9.68 and 19.60 mm, respectively.

NUMERICAL APPROACH

The flows in the supersonic separator are very complicated, including swirling flow and supersonic velocity. FLUENT software is the CFD solver of choice for complex flows ranging from incompressible (low subsonic) to mildly compressible (transonic) to highly compressible (supersonic and hypersonic) flows. Therefore, the commercially available FLUENT code was utilized here for the investigation.

Turbulence Model and Calculation Method

As a result of the high Reynolds number in the separator, the modeling of the turbulence plays a crucial role in the flow prediction. Turbulence models have been used for solving Reynolds-averaged Navier–Stokes (RANS) equations, because the direct simulation approach for turbulence is well beyond the capabilities of the computers. In the several modeling schemes, the *RNG k-ε* turbulence model is used here, because it provides an option to account for the effects of swirl or rotation by modifying the turbulent viscosity appropriately.

Mathematical calculation adopted *RNG k-ε* turbulence model, finite volume method, second-order upwind scheme and wall function applied in along-wall, while coupling numeration of velocity field and tress was based on SIMPLE.

Mesh Generation and Convergence

The finite difference, finite element and finite volume methods are usually used for integration of governing differential equations. In these approaches, the solution domain is subdivided into discrete elements or volume through computational grid in space. The mesh affects the speed, accuracy and convergence of the numerical calculation. The mesh can be body fitted structured hexahedral elements or unstructured tetrahedral elements, depending on geometry complexities. The finite volume methods have been used to discretize the partial differential equations of the model. The polyhedral meshes allow the flexibility of an unstructured mesh to be applied to a complex geometry. The swirling vanes are twisted, therefore, unstructured tetrahedral meshes are generated for the numerical computation here.

Accuracy increases with larger grids. However, grid independency studies have shown that larger grids do not necessarily influence the accuracy and the iteration steps for convergence of the solution in the case of the supersonic nozzle flow, while the mesh size is in a certain range [12]. In this simulation, the Mach number along the axis and total mass flow rates in inlet and outlet are selected for grid independence. When the Mach numbers are consistent with different grid numbers and mass flow rate error is less than $\pm 1.0 \times 10^{-4}$, grid independence can be verified. The grid independence is investigated with the mesh cells of 253 746, 612 501 and 1134 602, respectively. Even 253 746 grids provide a sufficient grid independency. However, for judging the shock wave position exactly, simulations have been performed using the 612 501 cells in the present study, as shown in Fig. 2.

Figure 2: Computational mesh for the supersonic separator.

In this simulation, the target parameters for convergence criterion are energy, continuity, velocity components, turbulent kinetic energy and turbulent dissipation rate. The solutions are considered converged when the dimensionless residuals drop below 1×10^{-6} for the energy, 1×10^{-4} for the continuity, velocity components, turbulent kinetic energy and turbulent dissipation rate, while simultaneously reaching stationary. The computation is stopped after these criteria are achieved.

Boundary Conditions

According to the flow characteristics of the supersonic compressible fluid, boundary conditions are imposed as follows: pressure boundary conditions for inlet and outlet of the supersonic separator, respectively. No-slip and adiabatic boundary conditions are specified for the walls.

In this computational study, Turbulent Intensity and Viscosity Ratio are chosen for the *RNG k-ε* model. The turbulence intensity at the core of a fully-developed duct flow can be estimated from the following formula derived from an empirical correlation for pipe flows:

$$I = \frac{u'}{u_{avg}} = 0.16(R_{eDH})^{-1/8}$$

$$(2)$$

where u', u_{avg} and $ReDH$ are the root-mean-square of the velocity fluctuations, the mean flow velocity and Reynolds number, respectively. The turbulent viscosity ratio, μ_t/μ, is directly proportional to the turbulent Reynolds number.

Typically, the turbulence parameters are set so that $1 < \mu_t/\mu < 10$.

Total pressure and total temperature at the separator inlet are set to be 6 atm and 300 K, respectively. The static pressure and stagnation temperature are chosen for outlet boundary conditions, which are 4.2 atm and 285 K, respectively. Turbulence intensity and viscosity ratio are assigned as turbulence parameters. Turbulence intensity and viscosity ratio are 0.048 and 1 for inlet, 0.025 and 6 for outlet, respectively.

Equation of State

It is assumed that the gas obeys the model equation of state which enables one to take into account the effect of super-compressibility in Eq. (3).

$$PV = mRTZ \qquad (3)$$

where P, V, m, R, T, Z are the gas pressure, volume, mass, gas constant, temperature, coefficient of super-compressibility, respectively. The coefficient of super-compressibility Z is a function of pressure and temperature. Absolutely, the coefficient of super compressibility Z is negligible by which Z is assumed to 1 at the low pressure conditions.

The mole composition of natural gas is as follows: 2.04% N_2, 0.45% CO_2, 0.03% H_2O, 91.36% CH_4, 3.63% C_2H_6, 1.44% C_3H_8, 0.26% i-C_4H_{10}, 0.46% n-C_4H_{10}, 0.17% i-C_5H_{12}, and 0.16% n-C_5H_{12}.

Comparison and Verification

Although the numerical solution methods are fully detailed in the above paragraphs, it is necessary to validate whether this methods could be used to calculate the complex swirling flow with supersonic velocity. This presentation will compare the numerical results with analytical and experimental data in [15]. The effects of swirl on the mass flow rate were analytical and experimental studied by Boerner et al. [15]. The dimensionless parameter S was employed to characterize the swirl strength.

$$S = \left(\frac{\int (v/a^*)dA}{A} \right)_{throat}$$ (4)

Where v and a^* are tangential and critical velocity.

Boerner et al. presented the results in terms of a dimensionless parameter M, which is the ratio of the mass flow rate in the presence of swirl to the mass flow rate for a corresponding non-swirling flow. The influences of swirling parameter S on the mass flow rate parameter M are shown in Fig. 3. The mass flow rate decreases with the higher swirl strength. The numerical results agreed with the experimental dates.

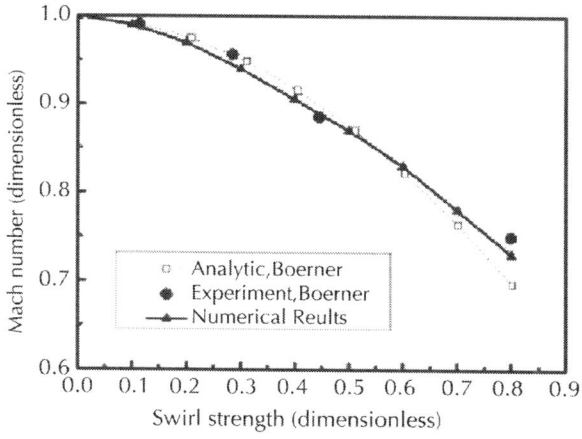

Figure 3: Effect of swirl on the mass flow through the nozzle.

RESULTS AND DISCUSSION

Static Temperature and Tangential Velocity

The flow field was simulated in the new supersonic swirling separator based on the above mentioned numerical methods. The static temperature and tangential velocity of natural gas in the supersonic separator are presented in Fig. 4.

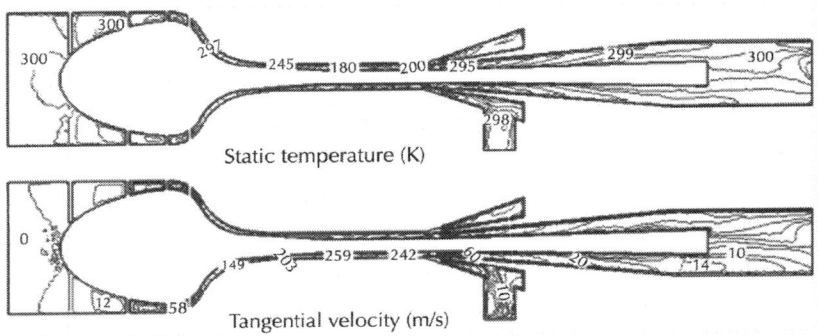

Figure 4: Static temperature and tangential velocity.

In the new supersonic swirling separator, low temperature is obtained as a result of the gas expansion in the divergent parts of the nozzle. The static temperature is about −93 °C at the nozzle exit. As a result of the low temperature, the nucleation of water and hydrocarbons occurs, followed by growth of liquid droplets. However, the static temperature increases slowly in the cyclonic separation section. The re-evaporation will decrease the separation efficiency.

The static vanes at the entrance swirl the natural gas flow into the nozzle. Judging by the principle of conservation of angular momentum, the swirl strength increases strongly in the annular nozzle. The strong swirls make a centrifugal field of about 720,000 × g (g is the acceleration of gravity), which centrifuges the liquid droplets onto the walls. Then, the gas liquid mixtures pass through the cyclonic separation section. The liquids are separated to the drainage pipe. The dry gas goes into the diffuser, where shock waves occur. As a result of the shock, the supersonic velocity is transferred to subsonic velocity. Therefore, pressure is recovered in the diffuser.

Radial Distribution

In the supersonic separator, the special features of the geometry induce the non-uniformly radial distribution of the main parameters of gas flow (velocity components, temperature, and so on). The disturbances significantly affect the flow characteristics of natural gas in the annular nozzle and the cyclonic separation section. The swirling flow is generated by the twisted vanes and a large vortex arises. The non-uniformity of radial distribution of gas velocity in the channel increases due to the convergence of the channel, which involves a redistribution of pressure and the emergence of return flows. Then, the vortex goes into the divergent and cyclonic separation sections from the convergent zone of the nozzle. Therefore, the droplets are centrifuged onto the walls.

It can be concluded that the presence of the strong swirling flow in the cyclonic separation section causes the nonuniform radial distribution of the main parameters of gas flow. Therefore, it is necessary to assess the effect of the radial distribution of the main parameters of gas flow on the process of separation. These sections investigate the dependence of natural gas dynamic parameters on the radius in the swirling separation cross section. The distribution of the gas temperature and the tangential components of the gas velocity are given in Fig. 5 and Fig. 6.

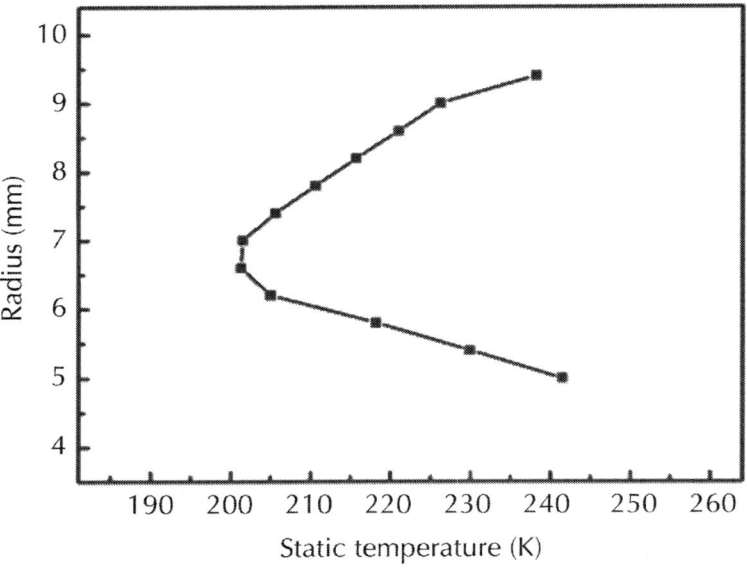

Figure 5: Static temperature.

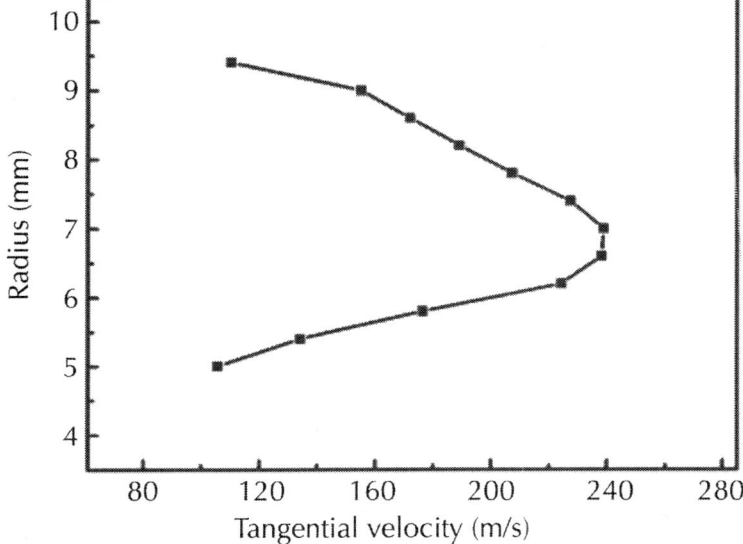

Figure 6: Tangential component of velocity.

The swirls are generated by the vanes and a large vortex arises in the subsonic part of the nozzle. Then, with further development of the natural gas swirling flow, the vortex goes deeper into the supersonic zone of the channel. The strong swirls induce the nonuniform radial distribution of the main parameters of gas flow at the cyclonic separation cross section. In the radial central region of the channel, the tangential component of velocity is distributed from 160 m/s to 240 m/s, on the radius from $r = 5.5$ mm to 9 mm at the swirling separation cross section, and the static temperature is radially distributed from 201 K to 225 K, which amounts to the range from −72 °C to −48 °C. That is, the centrifugal acceleration is up to the range from 280,000 × g to 820,000 × g, as well as the static temperature takes values from −72 °C to −48 °C, in the radial region from 5.5 mm to 9 mm of the cyclonic separation channel. Simultaneously, high gradients of gas dynamic parameters emerge in the vicinity of the channel walls. The presence of positive gradients of the velocity near the walls causes the mixing of the condensed components with the natural gas flow, which will decrease the separation efficiency. The high gradients of static temperature cause the re-evaporation of the condensed liquids. Therefore, the non-uniformly radial distribution of the gas dynamic parameters can impair the process of natural gas separation.

Effects of Shock Waves

A shock wave occurs due to an abrupt reduction in flow velocity that leads to changes in the flow from supersonic to subsonic. In the supersonic separator, the presence of the shock waves in the nozzle divergent and cyclonic separation sections prevents the natural gas expansion, which will destroy the low temperature for the condensation of some components. Therefore, the effects of the shock waves on the process of separation are assessed in these sections.

In the supersonic separator, the shock wave moves along the channel axis between the supersonic part of the nozzle ($z = 0.174$ m) and the entrance of the diffuser ($z = 0.242$ m), as shown in Fig.

7 and Fig. 8. Therefore, the extreme left (the worst) and extreme right (the best) positions of shock wave were selected for analysis, respectively. In advance, the pressure recovery coefficient   is defined as follow:

$$\gamma = \frac{p_{out}}{p_{in}}$$

(5)

where *pout* and *pin* are the outlet and inlet pressure of the supersonic separator, respectively.

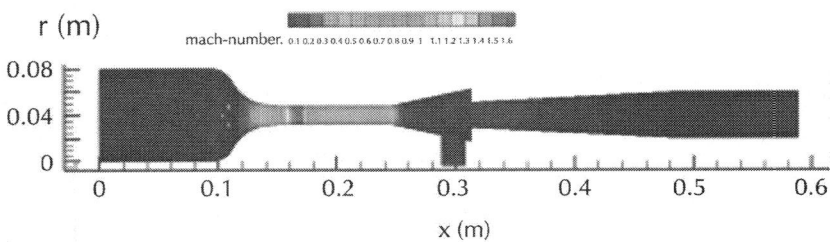

Figure 7: Shock wave in the nozzle.

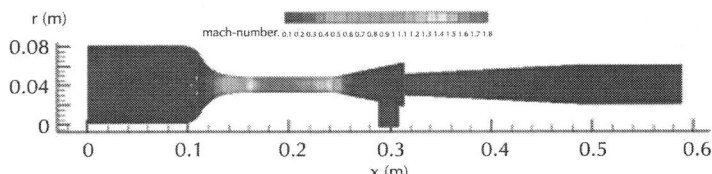

Figure 8: Shock wave in the diffuser.

In so doing, the pressure recovery coefficient at the channel outlet and inlet is 0.8 for Fig. 7, while 0.3 for Fig. 8. The pressure recovery coefficient has a great influence on the shock position. When the pressure recovery coefficient is increased gradually from 0.3 to 0.8, the shock wave moved into the expansion section of the nozzle from the diffuser.

If the shock wave is located almost immovably in the supersonic zone of the nozzle, though the upstream flow could expand to supersonic velocity, the flow downstream of the shock is subsonic, resulting in high temperature. When the shock wave takes the extreme left position ($z = 0.174$ m), the Mach number after the shock is distributed from $Ma = 0.8$ to $Ma = 0.56$, and the flow is subsonic in the cyclonic separation region. The gas average temperature at the cross section is distributed from $T = 266$ K to $T = 282$ K. As a result of the rapid rise of the temperature, the liquid particles, condensed by the low temperature, will re-evaporate into the gas, which leads to the failure of the gas and liquid separation of this device.

Only when the shock wave arrives at the entrance of the diffuser or goes into the diffuser, the liquid droplets can be centrifuged onto the wall and removed from the gas liquids mixtures. The effects of droplet re-evaporation could be neglected. When the shock wave takes the extreme right position ($z = 0.242$ m), the Mach number in the cyclonic separation section is distributed from $Ma = 1.8$ to $Ma = 1.62$, and the necessary conditions of the supersonic flow is valid. The gas average temperature in the cyclonic separation section is distributed from $T = 196$ K to $T = 220$ K, which is likewise favorable for the condensation of gas components.

The position of the shock wave determines the separation performance of the supersonic separator, and it could be affected by the pressure recovery coefficient. For this new design, the numerical results show that the shock wave reaches the entrance of the diffuser and the temperature of the flow upstream is 205 K, when the pressure recovery coefficient   is 0.78, which is the maximum to get the separator to work.

CONCLUSIONS

In this new supersonic swirling separator, gas expands to supersonic velocities with resulting in lower temperatures, which causes the condensation and nucleation of water and hydrocarbons. Meanwhile, the swirl strength is increased strongly due to the contraction of the channel. The very high centrifugal acceleration is in the order of 10^5 times the acceleration of gravity. Under the combined effect of the condensation and centrifugal field, water and heavy hydrocarbons can be condensed and separated from natural gas.

However, the main parameters of gas flow are non-uniformly distributed on radius due to the special features of the geometry. The tangential velocity and static temperature are uniformly distributed correspondingly in the radial central region of the cyclonic channel, while the non-uniformity of radial distribution causes high gradients of gas dynamic parameters in the vicinity of the channel walls.

The position of the shock wave significantly affects the process of separation. If a shock wave settles in the cyclonic separation section, it will result in high temperatures. This behavior of temperatures causes the re-evaporation of droplets, which hampers the separation of gas/liquids.

ACKNOWLEDGMENTS

This study is supported by National High Technology Research and Development Program of China("863", No. 2007AA09Z301) and National Major Science & Technology Specific Projects (No. 2008ZX05017-004).

REFERENCES

1.   S. Mokhatab, W.A. Poe, J.G. Speight, Handbook of Natural Gas Transmission and Processing, Gulf Professional Publishing, Burlington, MA, USA, 2006.

2.   M. Betting, H.D. Epsom, Supersonic separator gains market acceptance, World Oil 254 (2007) 197–200.

3.   M. Betting, T. Van Holten, J. Van Veen, US Patent, 6,524,368, February 25 (2003).

4. V.I. Alferov, L.A. Baguirov, V. Feygin, A. Arbatov, S. Imaev, L. Dmitriev, V.I. Rezunenko, US Patent, 6,372,019, April 16 (2003).

5. V.I. Alferov, L.A. Baguirov, L. Dmitriev, V. Feygin, S. Imaev, J.R. Lace, Supersonic nozzle efficiently separates natural gas components, Oil Gas J. 103 (2005) 53–58.

6. F. Okimoto, J.M. Brouwer, Supersonic gas conditioning, World Oil 223 (2002) 89–91.

7. H. Liu, Z. Liu, Y. Feng, K. Gu, T. Yan, Characteristic of a supersonic swirling dehydration system of natural gas, Chin. J. Chem. Eng. 13 (2005) 9–12.

8. W. Jiang, Z. Liu, H. Liu, J. Zhang, X. Zhang, Y. Feng, Two dimensional simulation and analysis of the flow in supersonic separator, J. Eng. Therm. 29 (2008) 2119–2121.

9. W. Jiang, Z. Liu, H. Liu, H. Pang, L. Bao, Influences of friction drag on spontaneous condensation in water vapor supersonic flows, Sci. China Ser. E – Technol. Sci. 52 (2009) 2653–2659.

10. W. Jiang, Z. Liu, H. Liu, J. Zhang, X. Zhang, Numerical simulation of twocomponent mixture in one-dimension supersonic separator, Acta Pet. Sinica Pet. Process 24 (2008) 697–701.

11. E. Jassim, M.A. Abdi, Y. Muzychka, Computational fluid dynamics study for flow of natural gas through high-pressure supersonic nozzles: part 1. Real gas effects and shockwave, Pet. Sci. Technol. 26 (2008) 1757–1772.

12. E. Jassim, M.A. Abdi, Y. Muzychka, Computational fluid dynamics study for flow of natural gas through high-pressure supersonic nozzles: part 2. Nozzle geometry and vorticity, Pet. Sci. Technol. 26 (2008) 1773–1785.

13. X.W. Cao, L. Chen, Z.H. Lin, Y.J. Dun, The nozzle used in supersonic swirling separators, Nat. Gas Ind. 27 (2007) 112–114.

14. A. Karimi, M.A. Abdi, Selective dehydration of high-pressure natural gas using supersonic nozzles, Chem. Eng. Process. 48 (2009) 560–568.

15.   C.J. Boerner, E.M. Sparrow, C.J. Scott, Compressible swirling flow through convergent-divergent nozzles, Heat Mass Transfer 5 (1972) 101–115.

Chapter 8

Selective Dehydration of High-pressure Natural Gas Using Supersonic Nozzles

Anahid Karimi and Majid Abedinzadegan Abdi

Faculty of Engineering and Applied Science, Memorial University of Newfoundland, St. John's, NL, A1B 3X5 Canada

ABSTRACT

Supersonic separators are proposed in this paper as a compact high-pressure processing system capable of selectively removing water from high-pressure natural gas streams without affecting the hydrocarbon content. A computer simulation linked to a thermodynamic property package is presented to predict the water removal efficiency and to compare the proposed system with conventional techniques. Some of the advantages of the proposed system include compactness, self-induced refrigeration, high gas velocity in the nozzle and low risk of hydrate deposition. Selective water removal with the proposed method can be achieved by controlling the design parameters: e.g., increasing the inlet pressure with constant temperature, increasing the inlet temperature with constant outlet pressure, and controlling the backpressure.

INTRODUCTION

The dwindling high quality crude oil reserves and increasing demand for natural gas has encouraged energy industries further towards the discovery of remote offshore reservoirs. Consequently, new technologies have to be developed to efficiently produce and transport stranded natural gas to consuming markets. Compactness of production systems is the most challenging design criteria for offshore applications. From the gas quality perspective, water vapour is the most common impurity in natural gas mixtures. The demand for natural gas has motivated the oil and gas industry to discover natural gas reservoirs in remote and harder to reach locations. The global need for less-carbon and potentially no-carbon content fuels (such as hydrogen) is motivated by environmental concerns. Natural gas is, at present, the only hydrocarbon energy source that will lead to major reductions in greenhouse gases and other pollutants. Natural gas, produced from the reservoir is not a single-component mixture, rather a mixture of hydrocarbons, which may include heavier-than-methane hydrocarbon constituents (C_2^+) or natural gas liquids (NGLs), reservoir water, and various impurities such as inert gases, carbon dioxide, and hydrogen sulphide. Natural gas needs to be processed before being used in the supply network. The impurities such as nitrogen, carbon dioxide, hydrogen sulphide, and heavy hydrocarbons can be removed in a central plant [1]. However, some other impurities such as sand and free water should be removed near the wellhead. Produced natural gas, in most cases, is in a supercritical dense phase. During natural gas processing it is likely that the water and the hydrocarbon components condense and form a liquid phase. This phase behaviour can be explained using the equilibrium phase diagrams known also as phase envelopes. The presence of water in natural gas decreases its heating value and if condensed cause major operational problems such as corrosion, excessive pressure can drop, hydrate formation and consequently the slug flow and reduction in gas transmission efficiency. The possibility of pipe obstruction due to the formation of hydrate within the flow lines is one of the most serious problems in the gas industry. The point at which gas hydrate forms and therefore becomes a source of trouble depends on gas pressure, temperature, and composition. Within the transportation system and at very high pressure of the gas, hydrate can form even at relatively high temperatures (close or above 20 °C). Therefore, it is important to assure that hydrate does not form as the gas is transported from the wellhead to a processing facility. Line heating, injection of hydrate inhibitors, and dehydration are commonly practiced to meet this requirement [2].

In processes such as transmission of gas in high-pressure pipelines and the gas storage in high-pressure containers for land or marine transportation of gas in compressed form, in certain specific pressure and temperature conditions, the presence of heavier hydrocarbons in natural gas is favourable [3]. As the heavier hydrocarbons (C_2^+) are introduced in the gas stream, the gas gravity increases and the compressibility factor decreases. Retaining the heavier hydrocarbons in the gas also eliminate the need for dewpointing and NGL recovery units. This may be considered as an advantage especially if the reservoir is in hard to reach environment, examples are arctic gas or gas in Deep Ocean or very cold environments where construction and running of large gas processing facilities may not be justified. The gas under such circumstances needs to be prepared to a "transportable" quality rather than a "pipeline" quality and therefore the removal of heavier hydrocarbons might not be an absolute necessity. Hydrate should however be handled properly no matter where the gas is produced from.

Natural gas can be dehydrated using the following different methods:

- Absorption using liquid desiccants.
- Adsorption using solid desiccants.

- Dehydration with calcium chloride.
- Dehydration by refrigeration.
- Dehydration by membrane permeation.
- Supersonic dehydration.

Extensive literature is available on common gas dehydration systems including solid and liquid desiccant and refrigeration-based systems [4] and [5]. Glycols are very good absorbers for water because the hydroxyl groups in glycols form similar associations with water molecules. The contact between a wet gas and glycol can be made in any gas–liquid contact device. Liquid desiccant systems are very established dehydration systems. They are relatively simple to operate and maintain and it is possible to automate them for unmanned operations [4] and [6]. This technology needs a large facility and due to the need for glycol, there is a possibility for some operational problems such as corrosion, foaming in contactor device, fouling of heat transfer surfaces, glycol contamination and loss.

Solid desiccant dehydration is also known as dry-bed dehydration. It uses a solid reagent to remove water. Adsorbents also known as desiccants are high capacity materials for water removal; examples include alumina, silica gels, and molecular sieves. Desiccants have limited capacity for water, become saturated soon, and therefore should be regenerated to restore their adsorptive capacity. The regeneration is usually accomplished by heating. Dry-bed dehydration is a semi-continuous process for which at least two parallel vessels filled with the adsorbent are required. In this arrangement, one vessel is adsorbing while other is regenerating [4], [5] and [6]. Solid anhydrous calcium chloride ($CaCl_2$) which forms various $CaCl_2$ hydrates when combined with water can be also used as desiccant to dehydrate natural gas. As water absorption continues, brine solution will be formed. In this unit calcium chloride pellets are placed in a fixed bed. The units might show poor performance under some conditions if $CaCl_2$ pellets bond together and form a solid bridge in the tower [4]. These units produce a waste stream that has to be taken care of appropriately.

Refrigeration through external vapour recompression is the simplest and most common process for natural gas dew point control (i.e., control of NGLs and water content of natural gas). In external or mechanical refrigeration systems the cooling is supplied by a vapour recompression cycle that typically uses propane or other common refrigerants as the working fluid. The refrigerant boils off and leaves the chillers as a saturated vapour [4] and [7]. If the gas inlet pressure is high enough, there will not be a need for external cooling and the expansion refrigeration that is known as low temperature extraction (LTX) or low temperature separation (LTS). This process applies the Joule–Thompson (JT) effect to reduce the gas temperature upon expansion in order to condense water and hydrocarbon and recover condensate with or without hydrate inhibition. A valve is used to throttle the high-pressure gas stream and generate the self-cooling, The JT effect in this process induces "self-refrigeration" as opposed to "external refrigeration" used in vapour recompression cycles discussed before. This technique requires a large pressure drop so it is used when a high-pressure gas is available. Turbine expansion can also be used for self-refrigeration instead of JT valves. In this case the lost energy could be substantially recovered by connecting the turbine shaft to a compressor [6].

Membranes have been successfully used to remove acid gases from natural gas. They have also been successfully used for dehydration of air. They are also being promoted by suppliers of membrane technologies for water removal [8]. They are relatively expensive (especially for large gas flow rates) and can be easily fouled by gas contaminants. They also need high pressure for efficient operation. However, they have a low-pressure drop through the process and do not need any chemical reagents. The installation and change of the membrane cartridges are relatively easy and the maintenance cost is low. The membranes' capability to remove water vapour is not selective and part of the gas is always wasted through co-permeation.

Most of the previously mentioned methods may have good dehydration performance but they have some disadvantages including the need for relatively large facilities, a considerable investment, complex mechanical work, and the possibility of having a negative impact on the environment. Supersonic gas processing systems were introduced to overcome some of the disadvantages of the alternative processes for dehydration [9]. The main part of a supersonic system is a supersonic (converging–diverging) nozzle. Supersonic nozzles are simple in design and do not contain any moving parts. In a supersonic nozzle both condensation (or solidification of hydrate) and separation occur at supersonic velocities, which leaves hydrate no time to deposit on the wall surfaces due to the short residence time and the high velocity of the fluid. Supersonic nozzles have been commercialized for dew point control applications [10]. The simplicity of this device makes it suitable for unmanned operation for underwater or remote gas production applications. As a result, it is claimed that the gas in this system can be dehydrated in a smaller, lighter, cheaper, more environmentally friendly, and less complex facility [9] and [11].

In a supersonic unit, the gas temperature is lowered based on gas expansion principles without the need of any refrigerant. The compactness of this design is a major advantage over traditional means of dehydration particularly for offshore applications. The gas velocity in this device is very high which prevents fouling or deposition of solids and ice. Refrigeration is self-induced therefore no heat is transferred through the walls and unlike external refrigeration systems, no inhibitor injection and inhibitor recovery system are necessary. The major drawback of this system is the pressure loss due to the expansion in the nozzle. The system is also very sensitive to variation of pressure and gas flow rate, therefore the turn-down ratio can also be limited. Most of the traditional means of dehydration remove water and hydrocarbon simultaneously and are not selective to any one chemical alone. At certain conditions of pressure and temperature, presence of heavy hydrocarbons (C_2^+) increases the gas gravity and reduces the compressibility factor, which results in an increase in the pipeline mass flow capacity [3]. Furthermore, the compactness and reliability of the process equipment are very important especially for offshore applications where the foot print area is at a premium. To remove water selectively natural gas should be kept in a single phase and hydrocarbon condensation should be avoided. This may require high pressures to keep the hydrocarbon in the supercritical state.

NOZZLE FLOW BEHAVIOUR

This paper focuses on predicting the flow behaviour and the shockwave location in a supersonic nozzle with a given dimension. The principles used in the analysis were discussed in our previous work [12] and [13]. Two software packages were linked to perform the mathematical analysis in this work namely, MATLAB (Version 7.0.4), a numerical computing environment and programming language and HYSYS (Version Aspen-HYSYS 2006), a process modeling and simulation software. Connecting these two software packages leads to a powerful simulation tool to study new processes [14] and helped this study to analyze various process alternatives.

As the stream pressure and temperature is reduced by gas expansion within the nozzle, the water vapour starts to condense to the aqueous phase. The condensed liquids (including possibly NGLs in case the pressure is below the cricondenbar) should be removed from the gas stream before the nozzle diverges downstream of the shockwave (see Fig. 1). Some gas leaves the nozzle together with the liquids, which should be separated and returned to the main gas leaving the nozzle at the exit. The suitable nozzle should be chosen such that the gas leaving at the exit contains the desired level of water and/or heavier hydrocarbons.

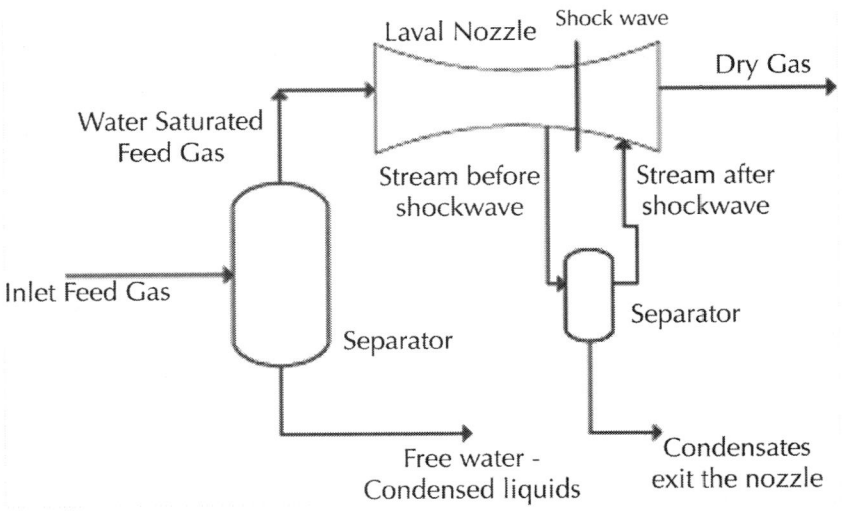

Figure 1: Schematic diagram of a supersonic dehydration unit.

A mathematical model based on MATLAB programming was developed to solve the governing equations of the system such as the continuity and momentum equations as well as the first and second laws of thermodynamics [12]. The Newton–Raphson technique was used to solve non-linear simultaneous equations [17]. Besides, an equation of state is needed to calculate the thermodynamic properties of the gas stream such as density, viscosity, enthalpy, and entropy (see Appendix A). The HYSYS process simulator was used to predict gas properties using commonly used thermodynamic models and fluid properties estimation techniques. The link between MATLAB and HYSYS gives the simulation model the capability of working with different EOSs. Input parameters to HYSYS are gas composition, inlet temperature, pressure and the flow rate. Free water is almost always present at well head conditions and it is therefore assumed that the produced gas is saturated before entering the process.

The flow through the nozzle is assumed to be supercritical, steady state, one-dimensional, isentropic, and compressible with no heat exchange.

Model Validation

The proposed model to simulate the flow behaviour in the supersonic nozzles was validated by the computational fluid dynamics (CFD) modeling using commercial CFD software (Fluent, Version 6.3.26)[12].

The nozzle geometry and other nozzle's parameters are shown in Table 1. The nozzle has a rectangular cross-section area and the inlet and outlet cross-section areas are 0.04 and 0.03 m², respectively.

Table 1: Nozzle geometry for validation with Fluent

Nozzle inlet diameter (m)	0.0400
Nozzle throat diameter (m)	0.0163
Nozzle outlet diameter (m)	0.0300
Nozzle converging length (m)	0.0527
Nozzle diverging length (m)	0.1473
Nozzle total length (m)	0.2000

To predict the flow behaviour in the supersonic nozzle with the proposed model in this work, pressure, temperature and the flow rate should be known as the inlet conditions. However, in CFD simulation, the inlet pressure and temperature are sufficient to predict the flow behaviour in the nozzle and the flow rate will be computed based on the inlet conditions. Fluent uses a rather complex k–e turbulent model and solves the governing partial differential equations using finite volume approach to predict the flow field behaviour. The proposed model in this work uses the simplified one-dimensional conservation laws and other related thermodynamic relations to predict the variation of various parameters along the nozzle. Table 2 lists the inlet pressure and temperature. For the model used in this work, the inlet flow rate is predefined at 67,330 kmole/h (300 kg/s) and the predicted flow rate with CFD modelling is the same.

Table 2: Nozzle inlet flow conditions for validation

Inlet temperature (°C)	18.5
Inlet pressure (kPa)	9250

The flow behaviour in the absence and presence of friction is predicted in both models. The results of this comparison with 75.67% inlet pressure recovery (backpressure of 7000 kPa) are shown in Fig. 2, Fig. 3, Fig. 4, Fig. 5 and Fig. 6.

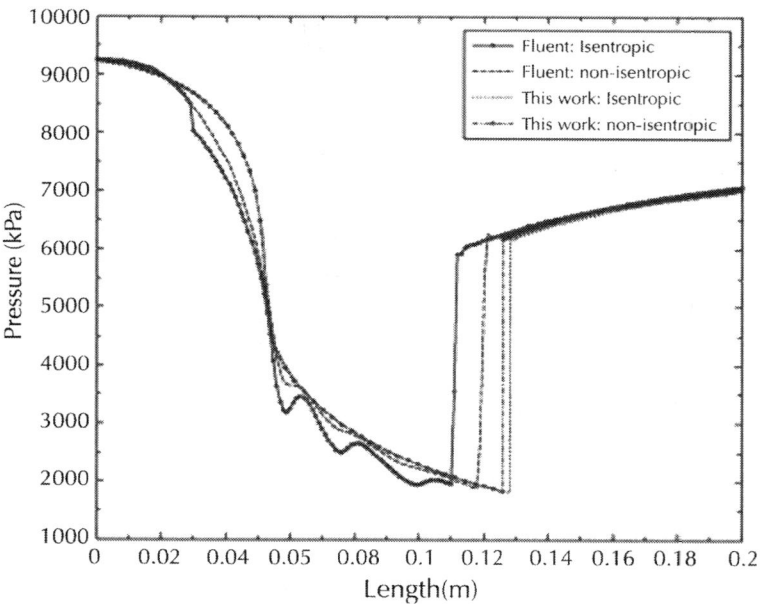

Figure 2: Pressure distributions and shockwave locations along the nozzle with 75.67% recovery of inlet pressure in the supersonic-CFD comparison study.

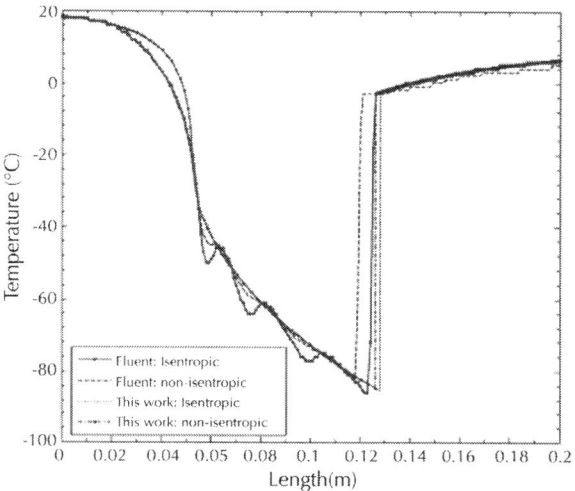

Figure 3: Temperature distributions and shock locations along the nozzle with 75.67% recovery of pressure inlet in the supersonic-CFD comparison study.

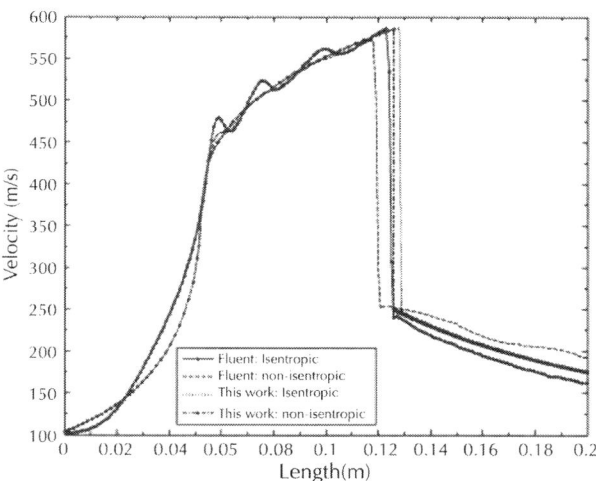

Figure 4: Velocity distributions and shock locations along the nozzle with 75.67% recovery of pressure inlet in the supersonic-CFD comparison study.

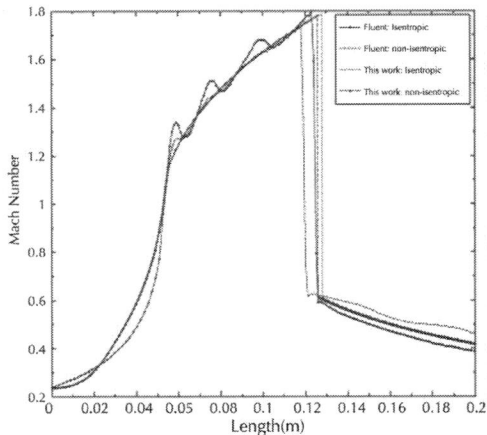

Figure 5: Mach number distributions and shock locations along the nozzle with 75.67% recovery of pressure inlet in the supersonic-CFD comparison study.

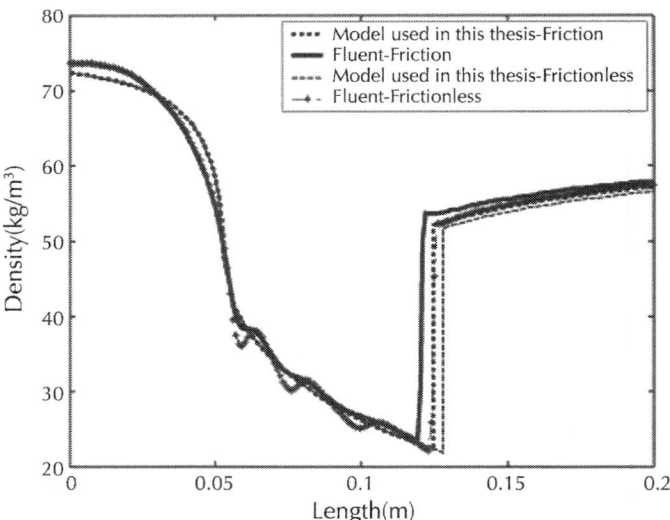

Figure 6: Density distributions and shock locations along the nozzle with 75.67% recovery of pressure inlet in the supersonic-CFD comparison study.

Both simulations showed that the flow behaviour along the nozzle is not significantly affected by the presence of friction. The friction however changes the location of shockwave when it happens inside the nozzle. The shockwave location in the frictionless flow is ahead of the frictional flow and towards the nozzle exit. Table 3 shows the shock locations in both simulations, in the cases where friction is present, and in the frictionless case as a percentage of the total length.

Table 3: Shock location predicted with the two simulators (% of total length)

	Frictionless	Friction	Deviation (%)
CFD modeling	61.5	59.5	3.25
Model used in this thesis	64	62.25	2.73

The flow behaviour predicted by the proposed method for supersonic separator agrees extremely well with the behaviour predicted by a complex computational modelling software (Fluent), giving rise to the accuracy of computational modelling and analyses preformed in this work.

The water content of the natural gas is a function of pressure and temperature and gas composition. The thermodynamic models used in this study (see Appendix A for the details) were employed to predict water content based on the equilibrium calculations. The results of water content calculations were compared with those obtained from McKetta Charts [4] and the accuracy was within ±5–10%. We could not claim that the result would also agree well with the actual nozzle performance until experimental tests are conducted. Such tests are beyond the scope of the current research work. The methodology however helped to investigate the efficiency of the supersonic nozzles for the application of interest in this study.

EXAMPLE STUDIES

Several example studies were run using the simulation model [12] to study the performance of supersonic nozzles for the high-pressure gas dehydration application. The properties of the "*Test Stream*", which is chosen as the working fluid of the nozzle in this study, are shown in Table 4 and Table 5.

Table 4: "Test Stream" gas composition

Gas components	Mole%
Methane	95.0
Ethane	4.0
Propane	1.0
Water (ppm)	230

Table 5: "Test Stream" inlet condition

Temperature (°C)	20
Pressure (kPa)	30,000
Molar flow rate (kmole/h)	5,000

The supersonic nozzle chosen based on these properties was set to choke the flow at the nozzle throat. Table 6 shows the geometry of the chosen nozzle.

Table 6: Fixed parameters used to design nozzle for example studies

Nozzle inlet diameter (m)	0.04
Convergence half angle	6.85
Divergence half angle	3.00
Throat diameter (m)	0.021

Convergence length (m)	0.082
Divergence length (m)	0.038
Exit diameter (m)	0.024

The effect of different parameters such as the inlet pressure, inlet temperature and flow rate on the behaviour of the chosen working fluid is discussed in this paper.

Effect of Inlet Pressure

The "Test Stream" (with no water content) at different pressures of 10, 30, 50, and 70,000 kPa was introduced to the nozzle. The stream capacity to hold water will decrease by the increase in inlet pressure as shown in Table 7.

Table 7: Inlet streams conditions for inlet-pressure-effect studies

Stream name	1	Test Stream	2	3
Temperature (°C)	19.72	19.85	19.86	19.87
Pressure (kPa)	10,000	30,000	50,000	70,000
Molar flow rate (kmole/h)	5,001	5,001	5,000	5,000
Water content (mg/m³)	264	174	150	109

Since the nozzle is chosen based on the properties of the "Test Stream", the same flow rate is either not enough or more than enough to achieve the sonic conditions at the throat for the other streams. Therefore, the inlet flow rates were adjusted for each stream condition. The flow capacity of the nozzle increases as pressure increases and will be 1421, 5000, 8016, and 10,544 for pressures of 10,000, 30,000, 50,000, and 70,000 kPa, respectively.

In this study, shockwaves occur in the nozzle under all conditions of pressure and temperature, when a 70% recovery of the inlet pressure is desired. As Fig. 7 shows, the decrease in inlet pressure will shift the shockwave location towards the nozzle exit.

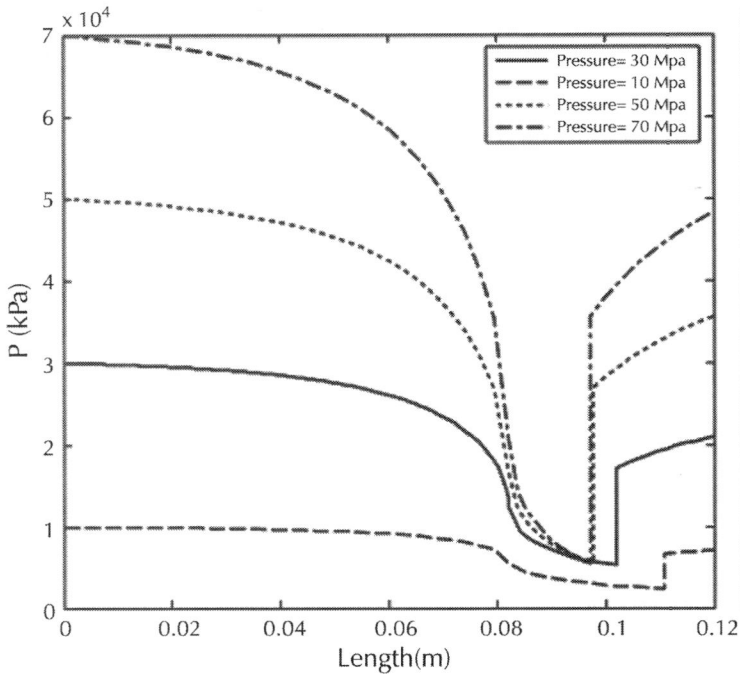

Figure 7: Pressure distribution in pressure-effect studies with 70% inlet pressure recovery.

Plotting the temperature variation with pressure for a nozzle in which 70% of the inlet pressure recovery illustrates that under the four gas conditions, the pressure–temperature lines remain in the dense phase region all the time. However, as the inlet pressure increases the temperature–pressure variation is more likely to stay in the dense supercritical phase of the phase envelope (see Fig. 8).

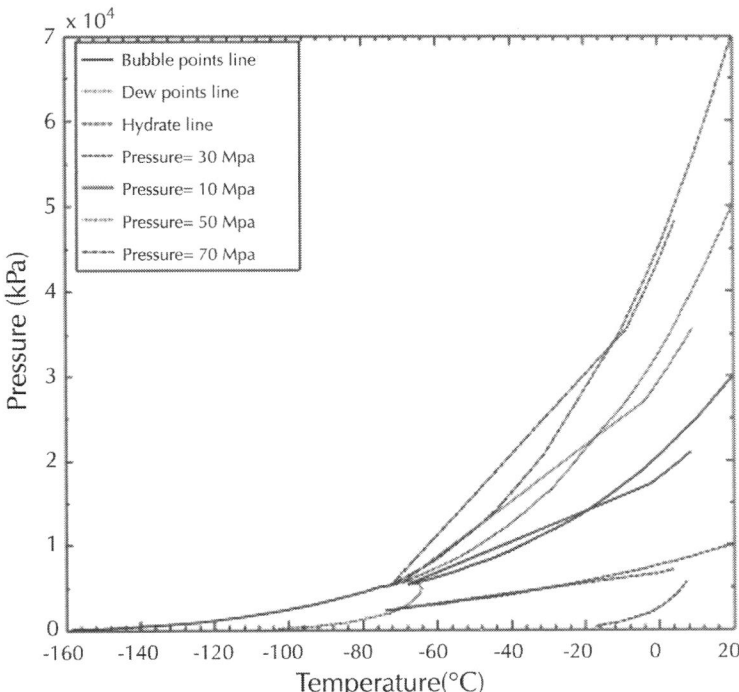

Figure 8: Phase envelope and pressure–temperature distributions for pressure-effect studies with 70% inlet pressure recovery.

Effect of Inlet Temperature

The "Test Stream" with different temperatures of 1, 20, 40, and 60 °C and equal pressure and flow rate were introduced to the nozzle. The capacity of gas to hold water increases as the temperature increases (see Table 8).

Table 8: Inlet condition of the streams for inlet-temperature-effect studies

Stream name	1	Test Stream	2	3
Temperature (°C)	1.0	19.8	39.6	59.0
Pressure (kPa)	30,000	30,000	30,000	30,000

Molar flow rate (kmole/h)	5,000	5,001	5,003	5,007
Water content (mg/m³)	60	174	465	1,085

If the 70% recovery of the inlet pressure is used as the criteria, the shockwave location varies depending on the inlet conditions (see Fig. 9). Increasing the inlet temperature from 1 to 60 °C, shifts the shockwave location towards the nozzle exit.

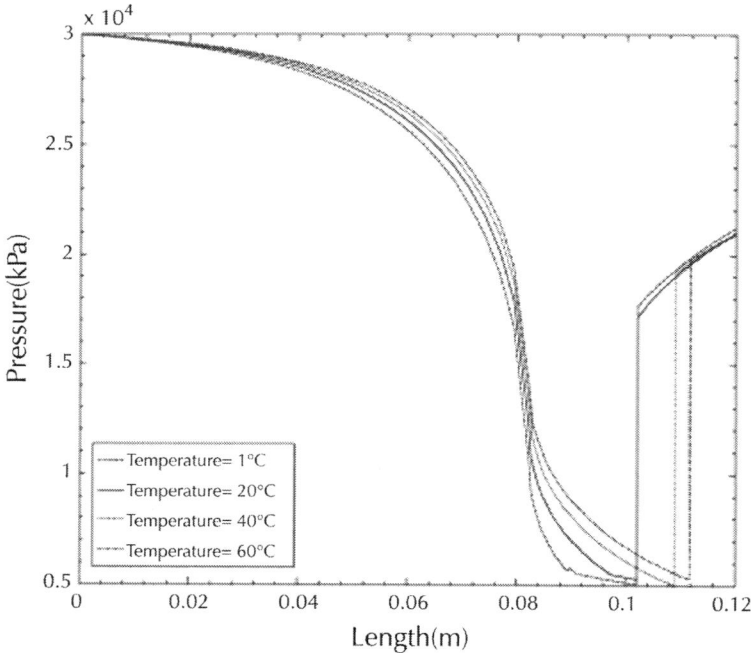

Figure 9: Pressure distribution and the shockwave location along a rated nozzle in temperature effect studies with 70% recovery of inlet pressure.

Fig. 10 illustrates that with 70% recovery of the inlet pressure, as the inlet temperature increases the amount of the liquid phase decreases and finally it can be seen that with the inlet conditions at 60 °C, the flow remains in a single phase all though the nozzle.

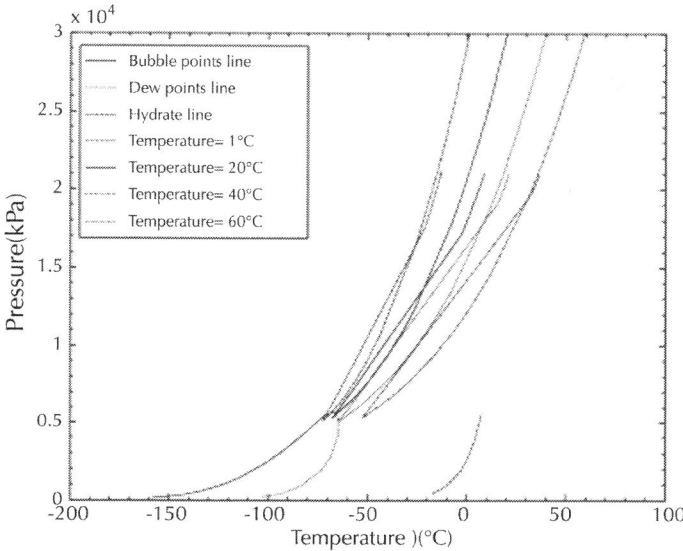

Figure 10: Phase envelope and pressure–temperature distributions along a rated nozzle in the temperature-effect studies with 70% inlet pressure recovery.

Effect of Flow Rate

In this study, we define a parameter called "design flow rate" for each stream condition and specified nozzle geometry at the design flow rate the critical conditions exist at the throat and the flow is choked. If the flow rate is lower than this specified flow rate, the Mach number will be less than unity at the throat of the nozzle and the flow will never be choked. In addition, if the flow rate is higher than the "design flow rate" then the throat Mach number should become greater than unity and this is an impossibility based on the supersonic nozzle theory. Fig. 11 and Fig. 12 show the Mach number and the pressure distributions of "*Test Stream*" at the different flow rates. For the flow rate equal to 5000 kmole/h, at which the nozzle was designed, the Mach number at the throat reaches unity. For the lower flow rates the stream is always remain subsonic (Fig. 6).

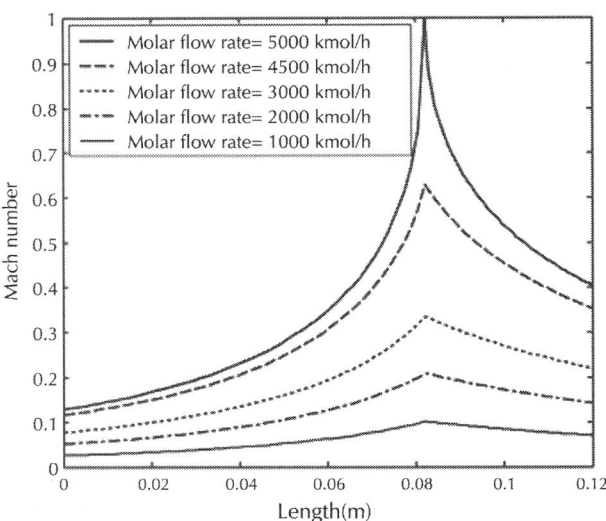

Figure 11: Mach number distribution along the nozzle with different flow rates.

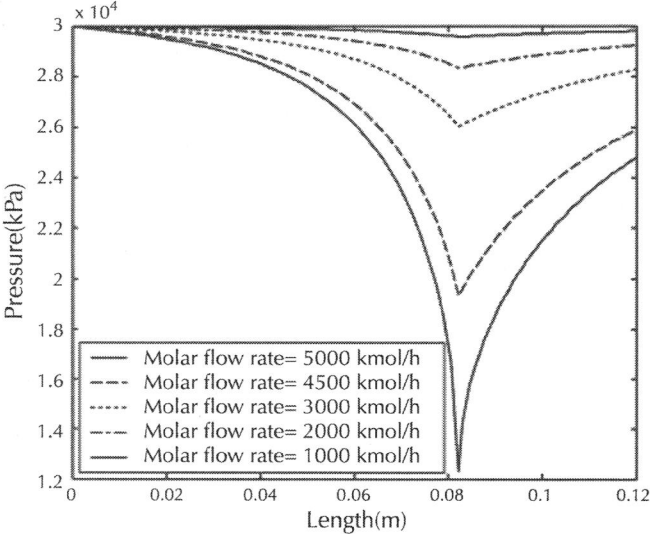

Figure 12: Pressure distribution along the nozzle with different flow rates.

As shown in Fig. 12, the pressure drop on both sides of the nozzle increases as the flow rate increases towards the "design flow rate".

Effect of Backpressure

The behaviour of the working fluid in the nozzle depends on the nozzle's back pressure (pressure at the nozzle exit). A supersonic nozzle might show two different behaviours after choking occurs at the throat. Pressure might recover after the throat or decrease even further. There are two pressures at the nozzle exit that need to be introduced here for further discussions namely "recovery pressure" and "design pressure" of the nozzle. At a certain ratio between the inlet pressure and the backpressure, the exit flow will be subsonic and most of the pressure can be recovered before any compression happens. This exit pressure is called "recovery pressure". If the flow is supersonic in the diverging part of the nozzle and no pressure recovery happens, the outlet nozzle pressure is called the "design pressure". For a nozzle of 0.12 m long and the "Test Stream" as a working fluid, the backpressure should not exceed 82.71% of the inlet pressure in order to choke the flow at the throat. This exit pressure is equal to the nozzle "recovery pressure". At this condition, the lowest pressure the nozzle can see at the end of its diverging part, "design pressure", is 14.84% of the inlet pressure (see Fig. 13). If the backpressure is between the nozzle "recovery pressure" and "design pressure", a shockwave happens either inside or outside of the diverging part of the nozzle. With the decrease in the backpressure, the shockwave location shifts towards the nozzle exit until the time the backpressure becomes equal to 48.5% of the inlet pressure where the shockwave happens at the nozzle exit. The shockwave happens outside of the nozzle if the backpressure decreases further down and falls between 48.50 and 14.84% of the inlet pressure (see Fig. 13).

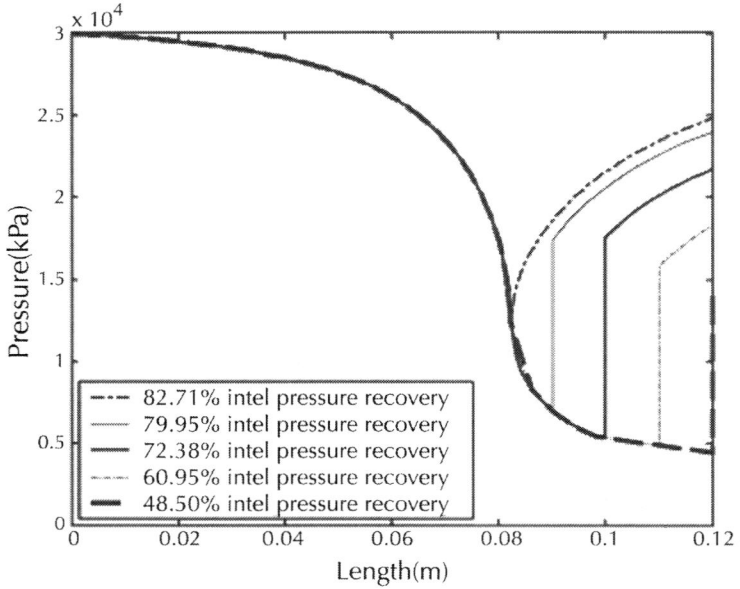

Figure 13: Pressure distribution along the nozzle for different backpressures.

Fig. 14, Fig. 15, Fig. 16 and Fig. 17 show the temperature–pressure variations and the locations of the lowest temperature with respect to the two-phase region. It is clear that by lowering the pressure recovery (higher pressure drops along the nozzle), the gas will expand more and pressure and temperature will be lower before the shockwave happens. If the selective water removal is desired, pressure and temperature before the shock should be kept outside of the two-phase region. As indicated in Fig. 9, Fig. 10, Fig. 11 and Fig. 12, higher backpressures cause the shockwave to happen earlier and less liquid hydrocarbons will form. Increasing the backpressure might however reduce the water removal efficiency. Table 7 lists the amount of water remained in the gas at different shockwave locations along the nozzle (Table 9).

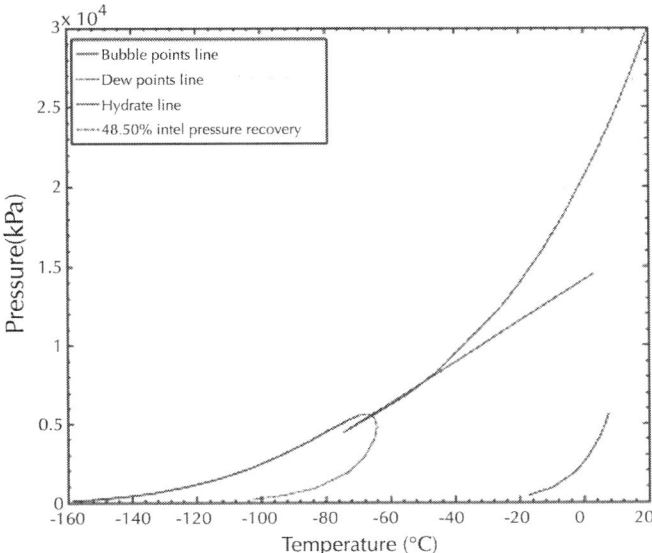

Figure 14: Phase envelope and pressure–temperature distributions with 48.5% inlet pressure recovery.

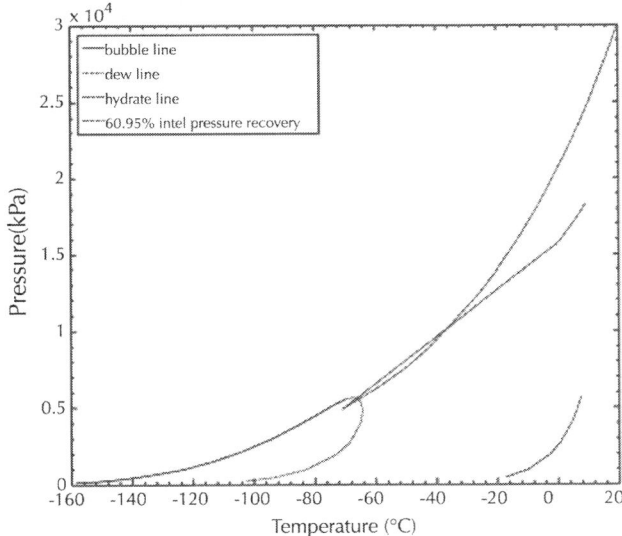

Figure 15: Phase envelope and pressure–temperature distributions with 60.95% inlet pressure recovery.

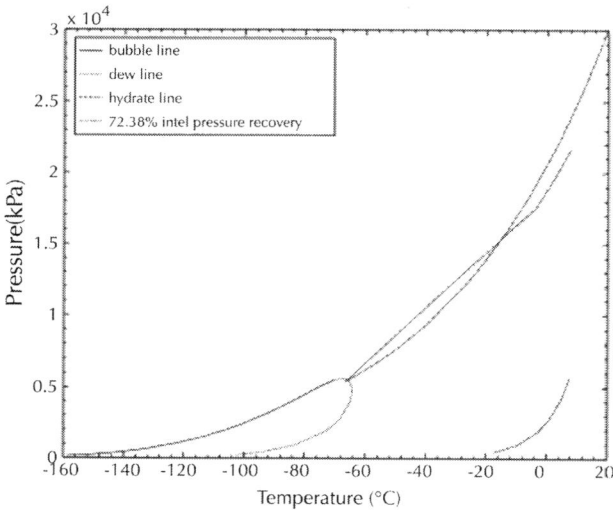

Figure 16: Phase envelope and pressure–temperature distributions with 73.38% inlet pressure recovery.

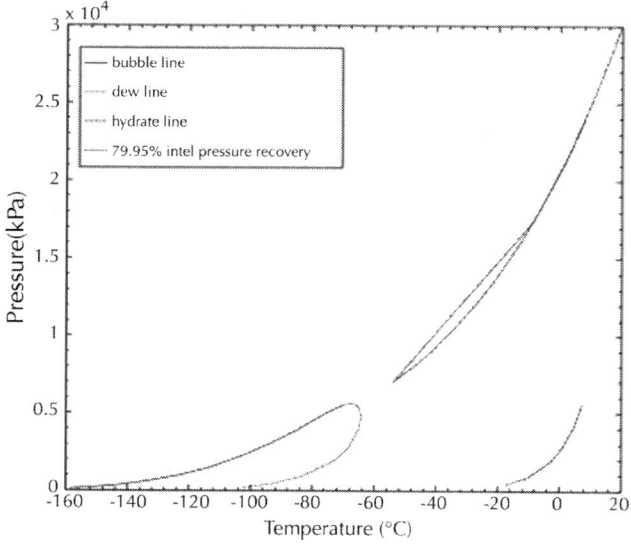

Figure 17: Phase envelope and pressure–temperature distributions with 79.95% inlet pressure recovery.

Table 9: Water content along the nozzle

Shockwave location (m)	0.082	0.090	0.100	0.110	0.120
Pressure recovery (% of inlet pressure)	82.7	80.0	72.4	61.0	48.5
Water remaining in vapour phase (% of initial water content)	6.00	0.50	0.10	0.06	0.04

CONCLUSIONS

A methodology was presented to simulate the performance of supersonic nozzles in removing water in supercritical conditions. The proposed model is capable of predicting the temperature and pressure profiles along the nozzle with a simpler algorithm therefore significantly lower computational time when compared with CFDs techniques. The proposed model shoed that the gas could potentially be dehydrated to very low water dew point temperatures using the commercially available supersonic nozzles.

APPENDIX A. THERMODYNAMIC MODELS (EQUATIONS OF STATE)

The most common EOSs used in the industry are Peng–Robinson (PR) and Soave–Redlich–Kwong (SRK) [15]. The general form of the equation is described in Eq. (A1).

$$P = \frac{RT}{V + \upsilon_1} - \frac{a_c\alpha(T)}{(V + \upsilon_2)(V + \upsilon_3)}$$
(A1)

where R is the gas constant; T is the temperature; P is pressure; V is molar volume; a_c is the attractive parameter at the critical temperature; α is the attractive parameter temperature dependence; υ_1,

v_2 and v_3 are volume correction parameters. The volume correction parameters are shown in Table A1 for some EOSs [16].

Table A1: Expression of volume correction parameters for generalized EOS in Eq. (A1)

EOS	Constant values		
	v_1	v_2	v_3
SRK	$-b$	0	B
PR	$-b$	$(1+\sqrt{2})b$	$(1-\sqrt{2})b$

Expression of volume correction parameters for generalized EOS in Eq. (A1)

In Table A1, the expressions of the volume correction parameters introduce parameter b, which is the molecular co-volume, and parameter c, which is a volume translation or volume shift parameter. The mathematical expression for a_c, α (T), b and c are given though Eqs. (A2), (A3), (A4), (A5), (A6) and (A7), and the values of the constants presented in these equations are indicated in Table A2 for the same EOS mentioned previously.

$$a_c = \frac{cte_1 R^2 T_C^2}{P_C} \tag{A2}$$

$$b = \frac{cte_2 RT_C}{P_C} \tag{A3}$$

$$\alpha(T) = \left[1 + m\left(1 - \sqrt{\frac{T}{T_c}}\right)\right]^2 \tag{A4}$$

$$m = m_1 + m_2\omega - m_3\omega^2 \tag{A5}$$

Where ω is the acentric factor.

$$c = \frac{C_1 RT_C(C_2 - Z_{RA})}{P_C} \tag{A6}$$

$$Z_{RA} = 0.29056 - 0.08775\omega \tag{A7}$$

Where Z_{RA} is the Rackett compressibility factor.

Table A2: Values of the EOS parameters

Constant	EOS	
	SRK	PR
cte_1	0.4274	0.45724
cte_2	0.08664	0.07780
m_1	0.48	0.37464
m_2	1.574	1.54226
m_3	0.176	0.26992
C_1	–	–
C_2	–	–

Since natural gas is a multi-component mixture, mixing rules are applied to calculate the parameters a, b, and c. The most applied mixing rules are the van der Waals one fluid prescription; however these are not applicable for nonpolar components of the mixture. The mathematical expressions of this mixing rule are indicated in Eqs. (A8), (A9) and (A10) for a, b, and c parameters.

$$a_c = \sum_{i=1}^{N} \sum_{j=1}^{N} x_i x_j \sqrt{a_i a_j}(1 - k_{ij})$$

(A8)

Where k_{ij} are the binary interaction parameter between components i and j, x is molar fraction, and N is total number of components.

$$b = \sum_{i=1}^{N} x_i b_i$$

(A9)

$$c = \sum_{i=1}^{N} x_i c_i$$

(A10)

REFERENCES

1.   B.D. Berger, K.E. Anderson, Gas Handling and Field Processing, Pennwell Corp., Tulsa, OK, 1980, pp. 59–75.

2.   L. Hengwei, L. Zhonggliang, F. Yongxun, G. Keyu, Y. Tingmin, Characteristic of a supersonic swirling dehydration system of natural gas, Chin. J. Chem. Eng. 13 (2005) 9–12.

3.   M. Mohitpour, H. Golshan, A. Murray, Pipeline Design and Construction, a Practical Approach, 2nd ed., ASME Press, New York, 2003, pp. 57–66.

4.   Gas Processors Suppliers Association, GPSA Engineering Data Book, 12th ed., Gas Processors Suppliers Association, Tulsa, OK, 2004, Section 16, pp. 1–13.

5.   F.S. Manning, R. Thompson, Oilfield Processing of Petroleum, Volume 1: Natural Gas, Pennwell Corp., Tulsa, OK, 1991, pp. 19–30.

6.   D. Ballard, Fundamentals of gas dehydration, in: Proceedings of the Gas Conditioning Conference, Oklahoma, Norman, and March 5–7, 1979.

7.   A.J. Kidnay, W.R. Parrish, Fundamentals of Natural Gas Processing, Taylor and Francis Group, LLC., 2006, pp. 9–15.

8.   K. Arnold, M. Stewart, Surface Production Operations: Volume 2; Design of Gas Handling Facilities, 2nd ed., Gulf Professional Publishing, Huston, TX, 1999, pp. 46–60.

9.   P. Schinkelshoek, H. Epsom, Supersonic gas conditioning introduction of the lower pressure drop twister, in: Presented in at the GPA 85th Annual Convention, Grapevine, TX, March 5–8, 2006.

10.   M. Betting, H. Epsom, Supersonic separator gains market acceptance high velocities make a unique separator and dew pointer, World Oil Mag. 228 (2007) 197–200.

11.   J.M. Brouwer, H.D. Epsom, Twister supersonic gas conditioning for unmanned platforms and subsea gas processing, in: Proceedings of the Offshore Europe Conference, 2003, pp.

219–225.

12.   A. Karimi, Selective Dehydration of High Pressure Natural Gas Using Supersonic Nozzles, M.Sc. thesis, Faculty of Engineering and Applied Science, Memorial University of Newfoundland, 2006.

13.   A. Karimi,M. Abedinzadegan Abdi, Selective removal of water from supercritical natural gas, SPE 100442, in: Proceedings of 2006 SPE Gas Technology Symposium: Mature Fields to New Frontiers, Calgary, Alberta, Canada, May 15–18, 2006, pp. 259–265.

14.   E.L. Beronich, K. Hawboldt, and M. Abedinzadegan Abdi, Linking a process simulator (HYSYS) with MATLAB, a powerful modeling tool for continuous process industry: a tutorial, in: Presented at the Annual NECEC Conference, Newfoundland, and November 9, 2005.

15.   T. Dustman, D. Bergman, J. Bullin, An Analysis and Prediction of Hydrocarbon Dew Points and Liquids in Gas Transmission Lines, GPA, 2006.

16.   K. Pedersen, P. Christensen, Phase Behaviour of Petroleum Reservoir Fluids, Taylor & Francis Group, 2007.

17.   R.B. Hollis, Real-Gas Flow Properties for NASA Langley Research Center Aerothermodynamics Facilities Complex Wind Tunnels, Langley Research Center, Hampton, Virginia, September, 1996.

Citations

CHAPTER 1

Snehanshu Pal and T. K. Kundu, "Theoretical Study of Hydrogen Bond Formation in Trimethylene Glycol-Water Complex," ISRN Physical Chemistry, vol. 2012, Article ID 570394, 12 pages, 2012. doi:10.5402/2012/570394.

CHAPTER 2

M.R. Rahimpour, M. Saidi, M. Seifi, Improvement of natural gas dehydration performance by optimization of operating conditions: A case study in Sarkhun gas processing plant, Journal of Natural

Gas Science and Engineering, Volume 15, November 2013, Pages 118-126, ISSN 1875-5100, http://dx.doi.org/10.1016/j. jngse.2013.10.001.

CHAPTER 3

Faheem A. Sheikh, Javier Macossay, Muzafar A. Kanjwal, Abdalla Abdal-hay, Mudasir A. Tantry, and Hern Kim, "Titanium Dioxide Nanofibers and Microparticles Containing Nickel Nanoparticles," ISRN Nanomaterials, vol. 2012, Article ID 816474, 8 pages, 2012. doi:10.5402/2012/816474.

CHAPTER 4

S.A. Al-Sobhi, A. Elkamel, Simulation and optimization of natural gas processing and production network consisting of LNG, GTL, and methanol facilities, Journal of Natural Gas Science and Engineering, Volume 23, March 2015, Pages 500-508, ISSN 1875-5100, http://dx.doi.org/10.1016/j.jngse.2015.02.023.

CHAPTER 5

Murilo D.M. Innocentini, Eduardo H. Tanabe, Monica L. Aguiar, José R. Coury, Filtration of gases at high pressures: Permeation behavior of fiber-based media used for natural gas cleaning, Chemical Engineering Science, Volume 74, 28 May 2012, Pages 38-48, ISSN 0009-2509, http://dx.doi.org/10.1016/j.ces.2012.01.050.

CHAPTER 6

Mohd Shariq Khan, Yus Donald Chaniago, Mesfin Getu, Moonyong Lee, Energy saving opportunities in integrated NGL/LNG schemes exploiting: Thermal-coupling common-utilities and process knowledge, Chemical Engineering and Processing: Process Intensification, Volume 82, August 2014, Pages 54-64, ISSN 0255-2701, http://dx.doi.org/10.1016/j.cep.2014.06.001.

CHAPTER 7

Chuang Wen, Xuewen Cao, Yan Yang, Swirling flow of natural gas in supersonic separators, Chemical Engineering and Processing: Process Intensification, Volume 50, Issue 7, July 2011, Pages 644-649, ISSN 0255-2701, doi.org/10.1016/j.cep.2011.03.008.

CHAPTER 8

Anahid Karimi, Majid Abedinzadegan Abdi, Selective dehydration of high-pressure natural gas using supersonic nozzles, Chemical Engineering and Processing: Process Intensification, Volume 48, Issue 1, January 2009, Pages 560-568, ISSN 0255-2701,.doi.org/10.1016/j.cep.2008.09.002.

Index